Energy and Environmental Outlook for South Asia

Energy and Environmental Outlook for South Asia

Edited by

Muhammad Asif

CRC Press

Taylor & Francis Group

Boca Raton London New York

CRC Press is an imprint of the
Taylor & Francis Group, an **informa** business

First edition published 2021
by CRC Press
6000 Broken Sound Parkway NW, Suite 300, Boca Raton, FL 33487-2742
and by CRC Press
2 Park Square, Milton Park, Abingdon, Oxon, OX14 4RN

Library of Congress Cataloging-in-Publication Data

Names: Asif, Muhammad, editor.
Title: Energy and environmental outlook for South Asia / edited by Muhammad Asif.
Description: First edition. | Boca Raton, FL : CRC Press/Taylor & Francis
    Group, LLC, 2021. | Includes bibliographical references and index.
Identifiers: LCCN 2020035655 (print) | LCCN 2020035656 (ebook) | ISBN
    9780367673437 (hardback) | ISBN 9781003131878 (ebook)
Subjects: LCSH: Energy policy–South Asia. | Power resources–South Asia. |
    South Asia–Environmental conditions.
Classification: LCC HD9502.S613 E53 2021 (print) | LCC HD9502.S613
    (ebook) | DDC 333.790954–dc23
LC record available at https://lccn.loc.gov/2020035655
LC ebook record available at https://lccn.loc.gov/2020035656

ISBN: 978-0-367-67343-7 (hbk)
ISBN: 978-1-003-13187-8 (ebk)

Typeset in Times LT Std
by KnowledgeWorks Global Ltd.

Dedication

To my son Ibraheem

Contents

Forewords..ix
Preface.. xiii
Editor xv
Contributors xvii

Chapter 1 Introduction 1
Muhammad Asif

Chapter 2 Afghanistan's Energy and Environmental Scenario 17
Najib Rahman Sabory, Mir Sayed Shah Danish, and
Tomonobu Senjyu

Chapter 3 An Overview of Energy Scenario in Bangladesh: Current
Status, Potentials, Challenges and Future Directions 39
Imran Khan, Sujan Chowdhur, and Zobaidul Kabir

Chapter 4 100% Renewable Energy Strategy: Bhutan's Energy and
Environmental Perspective.. 63
Hari Kumar Suberi

Chapter 5 Sustainable Energy and Environmental Outlook: Indian
Perspective 97
Asha Pandey

Chapter 6 Energy and Environmental Development in Maldives 123
Fathmath Shadiya

Chapter 7 Energy Security in the Context of Nepal 141
Tri Ratna Bajracharya and Shova Darlamee

Chapter 8 Pakistan's Energy Transformation Pathway and Environmental
Sustainability 175
Naila Saleh

Chapter 9 Energy, Environment and Sustainable Development Futures
in Sri Lanka.. 195

Ajith de Alwis and Maksud Bekchanov

Chapter 10 South Asian Dual Challenges: Energy and Environment................ 217

John Andrew Howe and Kankana Dubey

Chapter 11 Conclusion 253

Muhammad Asif

Index.. 265

Forewords

FOREWORD 1

This first book on energy and environment in South Asia is both timely and highly relevant. The distinguished editor, Dr. Asif, and talented chapter authors provide detailed coverage of all regional countries.

Balanced treatment of the three key dimensions of the sustainable development triangle (environment, economy and society) is important for the whole world, especially South Asia. Energy is intimately connected to environment via society and economy. Examining this nexus, the book shows how energy, socioeconomic and ecological systems have co-evolved to a critical point in the region. Unless future human activities are better oriented toward sustainability, especially sustainable energy development (SED), the living conditions of billions will be at risk. The authors provide valuable guidance regarding the way forward.

SED seeks to harness energy resources to make human development more sustainable, while harmonizing economic, social and environmental dimensions. First, the growth of modern economies depends on energy services such as heating, refrigeration, cooking, lighting, communications, and motive power. Over two billion people cannot access affordable energy services, impeding opportunities for economic development and improved living standards. Second, energy is a basic social need affecting human well-being. Wide disparities in access to affordable commercial energy and energy services in both urban and rural areas undermine social progress. Access to decentralized, small-scale energy technologies is important for poverty alleviation. Women and children, who are relatively more dependent on traditional fuels, suffer disproportionately. Third, there are many environmental links (mainly negative), since energy production and use continue to be a primary source of local, transnational and global pollution. Specific impacts include water contamination; land degradation; marine and coastal pollution; ecosystems destruction and loss of biodiversity; damage to health, structures and natural systems from SO_x, NO_x and particulates that degrade air quality; and finally, climate change driven by greenhouse gas (GHG) emissions that will worsen all other problems.

The socioeconomic impacts of climate change will be severe in South Asia. During the past three decades, 1.7 billion people were affected by over 1,000 climate-induced disasters, which caused almost $130 billion in damages. By 2030, climate change could drive over 60 million people in South Asia into extreme poverty, while floods alone would inflict $215 billion of damages annually. Adaptation to reduce climate change vulnerability must have high regional priority.

The balanced inclusive green growth (BIGG) path offers an overall framework to address all sustainable development issues in an integrated manner, and the 17 sustainable development goals (SDG) provide a practical implementation mechanism, after appropriately localizing to specific South Asian countries. Energy for SD will play a key role, requiring (a) more efficient use of energy, especially in end use – buildings, transportation and production processes; (b) increased reliance on renewable energy sources; and (c) accelerated development and deployment of new energy technologies.

Strategies and policies for globally achieving SED include better governance; encouraging greater international cooperation in areas such as technology, harmonization of environmental taxes and emissions trading, and energy efficiency standards for equipment and products; adopting mechanisms to increase access to energy services through modern fuels and electricity; enabling all stakeholders to participate in energy decisions; advancing innovation throughout the innovation chain; encouraging competitiveness in energy markets to reduce energy services costs; cost-based prices, especially ending subsidies for fossil fuels and nuclear power; internalizing external environmental and health impacts; encouraging greater energy efficiency; and developing and diffusing new technologies widely. Innovation, agility and sustainability in providing energy services will be even more important to face many uncertainties, as stressed socioeconomic systems undergo transformation after COVID-19.

This clearly written book will explore a full range of such options and help to build a better future for South Asians.

Prof. Mohan Munasinghe
Chairman, Munasinghe Institute for Development (MIND) and MIND Group
Honorary Senior Advisor to the Government of Sri Lanka
Vice Chair, Intergovernmental Panel on Climate Change (IPCC-AR4)
that shared the 2007 Nobel Peace Prize

FOREWORD 2

Energy and environment are fundamental cornerstones of sustainable development. The availability of sufficient, reliable and affordable energy has become almost a prerequisite to the socio-economic prosperity of a society. The United Nations also regards the energy and environmental well-being critical for sustainable development as is evident from the sustainable development goals (SDGs). Given the challenges, the global energy scenario is going through a paradigm shift in terms of technological and policy developments. Energy solutions and technologies such as renewables, electric mobility and fuel cells are defining the future of the world energy scenario. Especially renewable energy and electric mobility are already outpacing the conventional energy solutions and gasoline-based transportation respectively.

South Asia is the most populous developing region in the world. Despite being rich in terms of natural resources and capable manpower, its socio-economic indicators are far from satisfactory. Not surprisingly, South Asia's energy and environmental outlook also face serious challenges. Energy security, especially in terms of availability and affordability, is a major concern for the overwhelming majority of the region's population. Climate change is also posing threats requiring an urgent and meaningful response. Fast-melting Himalayan glaciers. Disruptions in water supplies from these rivers are bound to exacerbate the food and water security situations. Rising sea levels are posing major challenges for the region especially for countries such as Maldives and Bangladesh. Other weather-related calamities such as flooding, droughts and cyclones are also becoming more frequent and intense.

This book is an excellent contribution to South Asia's energy and environmental outlook. Besides providing a regional perspective, it presents the energy and environmental scenarios of all South Asian countries, also highlighting challenges faced and the potential solutions. The book has rightly debated that the broader progress and development in South Asia – integrally linked with its energy and environmental outlook – is being hampered by the geopolitical disputes in the region. South Asian countries need to cooperate in the field of energy and environment, especially in the fight against climate change. In the wake of the COVID-19 crisis, the book is a timely and valuable addition to the debate on post-pandemic recovery. As the Editor, Dr. Asif implies in the conclusions, the world needs to 'build back better' by exploring a more resilient lifestyle based on sustainable technologies and practices.

Atta-ur-Rahman
Fellow of Royal Society (London)
Honorary Life Fellow, Kings College, Cambridge University
Professor Emeritus, International Centre for Chemical
and Biological Sciences, Pakistan
Former Chairman, Higher Education Commission, Pakistan

FOREWORD 3

Energy insecurity and environmental degradation are two of the compelling sustainable development challenges faced by the world in general and the developing countries in particular. South Asian countries are experiencing energy- and environment-related concerns ranging from lack of access to reliable, quality and cost-effective energy fueling poverty and dependence on imported fossil fuels. Climate change implications such as rising sea level, melting of Himalayan glaciers and extreme weather events are posing serious threats to the millions of people living in South Asian countries. Better understanding and appropriate actions are needed to address this growing set of energy and environmental challenges to ensure socio-economic and technological advancement of these countries. The book provides a comprehensive account of South Asia's energy and environmental outlook. It incorporates a wealth of data and information on each country in the region. It is an excellent and commendable academic and research contribution on the subject beneficial to broader stakeholders including developmental and public sector entities involved in the area of energy and environment.

Barsha Man Pun Ananta
Minister Energy, Water Resources and Irrigation
Nepal

FOREWORD 4

Bhutan's energy mix is dominated by hydropower and biomass, the former contributing to over 95% of power generation. While the country has made good progress in electrification, its energy scenario faces the challenges of environmental emissions associated with the large-scale use of biomass, and financial burden as a result of fossil fuel imports. Besides hydropower and biomass, Bhutan has significant potential for solar energy and wind power generation. To satisfy the growing energy demand, it needs to tap other renewable resources including solar energy, wind power, modern bioenergy and small-scale hydropower. These resources can diversify the energy supply, enhance energy security and contribute toward the country's commitment to remaining carbon neutral.

While the country has consistently maintained the highest environmental conservation standards, it is not free from the global challenges posed by climate change. The Himalayan glaciers and river systems are vital for the livelihood of not only Bhutanese but also to the hundreds of millions of people living downstream. Climate change threatens this perennial source of livelihood and calls for further deepening of mutually beneficial collaboration in South Asia.

The book *Energy and Environmental Outlook for South Asia* provides valuable insight into the energy and environmental scenario of Bhutan as well as other countries in the region. And its publication will further contribute to better understanding of the energy and environmental challenges we face in South Asia and highlight the need to increase cooperation in these areas. Furthermore, this book is timely when governments are contemplating the role of renewable energy resources in the context of their respective climate change mitigation efforts.

Loknath Sharma
Minister of Economic Affairs
Bhutan

FOREWORD 5 (Web content only) Visit https://www.routledge.com/Energy-and-Environmental-Outlook-for-South-Asia/Asif/p/book/9780367673437

FOREWORD 6 (Web content only)

FOREWORD 7 (Web content only)

Preface

South Asia, home to almost a quarter of the global population, is one of the most important regions in the world for its large manpower base, vast natural resources, large business and economic markets, important geographical position, vibrant culture, and rich history. Despite this, South Asia has some disappointing statistics from the energy and environmental perspectives. Over 350 million people in the region lack access to electricity, and more than 1 billion people rely on crude biomass to meet their cooking needs. Ironically, South Asia has only 5% of the global electricity generation capacity to fulfill the needs of almost a quarter of the world's population. Unsurprisingly then, the per capita electricity consumption in the region is less than a quarter of the world average. The majority of those connected to the grid face reliability issues such as prolonged power outages, load shedding, and low voltages. In terms of quality of electricity supplies, India, Pakistan, and Bangladesh – the three large nations in the region – have a standing of 108, 99, and 68, respectively, out of 140 countries in the world. Rising sea levels, melting glaciers and other extreme weather-related calamities such as frequent flooding and droughts add greatly to the serious challenges to the region; low-lying areas of the Maldives and Bangladesh are now facing existential threats. During 1999–2018, Pakistan, Bangladesh, and Nepal were the 5th, 7th, and 9th, respectively, most affected countries in the world from extreme weather-related natural disasters. Climate change further aggravates the food and water security crisis in the region. In 2019, four out of five countries in the world with the worst levels of the particulate matter 2.5 (PM2.5) were noted in South Asia, while 27 of the 30 most polluted cities were located in the region, with India alone having 21 of them.

Geopolitically, South Asia is one of the most unstable regions in the world. Wide-ranging inter-country conflicts, especially territorial and water disputes, have frequently plagued the region, even resulting in wars, civil wars, and proxy wars. While their unsettled disputes have long made India-Pakistan a nuclear flashpoint, 2020 is witnessing India and China tangled in a similar situation. Notwithstanding a high level of poverty in the region—with almost half of the global below-poverty level population living here—the countries in the region have spent more than US$80 billion on their military affairs in 2019. The rivalries in South Asia are vividly undermining the energy and environmental outlook of the region.

The issues above are critical as South Asian countries have their socio-economic prosperity, techno-economic advancements, and geopolitical stability intrinsically linked to their energy and environmental security. There is a tremendous potential for these countries to cooperate in the field of energy, and benefit from the diverse range of resources, from within the region and beyond, to address their energy and environmental challenges. COVID-19 is a stark reminder for the governments and other stakeholders of South Asian countries to reflect and revamp their priorities. It is time to envision a future based on commonalities and cooperation, instead of hostilities to (re)build a more resilient lifestyle based on sustainable technologies and practices. Indeed, human interaction with nature and natural resources needs to be rearranged.

With the above backdrop, this book presents the case for South Asia's energy and environmental scenario improvement which is imperative not only for the region but also for the rest of the world. To maintain uniformity and focus in terms of the scope and the structure of the book, the energy and environmental outlook of each of the South Asian countries as well as of the region is discussed in a dedicated chapter. The fact that all country-specific chapters are contributed by experts from their respective countries makes it a unique scholarship. Contributors have thus diligently incorporated and reflected on the local context and knowledge along with passion in their analysis. All contributors show due proficiency and focus in terms of discussing the energy and environmental scenarios of their respective countries and the region. The book is a sort of wake-up call as well as a motivation for South Asian countries to act before it is too late.

Editor

Dr. Muhammad Asif is Associate Professor at the Glasgow Caledonian University, Scotland. His research interests include renewable energy, energy management, energy policy, sustainable buildings, life-cycle analysis, and sustainable development. He is the author and/or editor of three books and more than 100 peer-reviewed research papers, book chapters and reports, and has presented his research at more than 50 international conferences and symposia.

Contributors

Ajith de Alwis
Department of Chemical and Process Engineering, University of Moratuwa, Moratuwa, Sri Lanka

Muhammad Asif
Glasgow Caledonian University, UK

Tri Ratna Bajracharya
Department of Mechanical Engineering, Pulchowk Campus, Institute of Engineering, Tribhuvan University, Nepal

Maksud Bekchanov
Center for Development Research (ZEF), Bonn University, Bonn, Germany and Center for Earth System Research and Sustainability (CEN), Hamburg University, Hamburg, Germany

Sujan Chowdhury
Department of Chemical Engineering, Jashore University of Science and Technology, Jashore-7408, Bangladesh

Shova Darlamee
Department of Mechanical Engineering, Pulchowk Campus, Institute of Engineering, Tribhuvan University, Nepal

Kankana Dubey
Centre for Energy Policy, University of Strathclyde, UK

John Andrew Howe, PhD
Independent Researcher, UK

Zobaidul Kabir
School of Environmental and Life Sciences, University of Newcastle, Australia

Imran Khan
Department of Electrical and Electronic Engineering, Jashore University of Science and Technology, Jashore-7408, Bangladesh

Asha Pandey
Department of Environmental Science, School of Vocational Studies and Applied Sciences, Gautam Buddha University, Uttar Pradesh, 201308, India

Najib Rahman Sabory
Energy Department, Kabul University, Jamal Mina, Kabul, Afghanistan

Mir Sayed Shah Danish
Energy Department, Kabul University, Jamal Mina, Kabul, Afghanistan

Tomonobu Senjyu
Department of Electrical and Electronics Engineering, University of the Ryukus

Naila Saleh
Institute of Policy Studies, Pakistan

Fathmath Shadiya
The Maldives National University

Hari Kumar Suberi
Trier University of applied sciences Umwelt-Campus Birkenfeld, Germany; Freelance Energy Consultant in Bhutan

1 Introduction

Muhammad Asif

CONTENTS

1.1 Energy, Environment, and Sustainable Development 1
1.2 South Asia in Perspective 3
1.3 Energy and Environmental Challenges 6
1.4 Regional Geopolitics and Conflicts 10
    1.4.1 Territorial Disputes 11
    1.4.2 Water Disputes 11
1.5 Structure of the Book... 13
References 14

1.1 ENERGY, ENVIRONMENT, AND SUSTAINABLE DEVELOPMENT

Energy is a precious commodity that goes through a wide range of flows and transformations that are pivotal for the existence of human life on the planet. Increasingly extensive and efficient utilization of energy has played a critical role in the evolution of societies, especially in the post-industrial revolution era. Energy has become a prerequisite for almost all facets of life, i.e., agriculture, industry, mobility, education, health, and trade and commerce. The provision of adequate and affordable energy services is crucial to sustain a modern lifestyle, ensure development and eradicate poverty. The per capita energy consumption is an index used to measure the socioeconomic prosperity in any society. The United Nation's Human Development Index (HDI) has a strong relationship with energy prosperity (Asif 2011; BP 2019). A correlation between electricity consumption per head and economic well-being in a range of countries, for example, has been shown in Figure 1.1. It can be observed that an increase in electricity consumption up to 4,000 kWh/capita has a strong relationship with human welfare.

The global demand for energy is experiencing a rapid growth. According to the Energy Information Administration (EIA 2019), the world energy requirement will grow by 50% by 2050. This growth is mostly expected to come from the developing world as the countries outside the Organization for Economic Cooperation and Development (OECD) are likely to account for over 60% of this growth in energy demand (EIA 2017). The rapid growth in energy demand in developing countries is driven by factors such as burgeoning population, urbanization, modernization, and economic and infrastructure development. By the end of 2019, the world population grew to more than 7.7 billion (UN 2019). In 2008 for the first time in history nearly half of the global population lived in urban areas. It is estimated that by 2050 approximately 68% of the population will be living in urban areas. Projections also

1

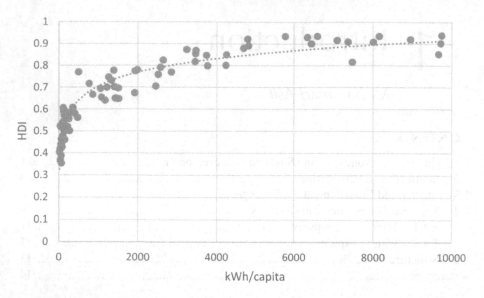

FIGURE 1.1 Relationship between electricity use and HDI of a wide range of countries.

suggest that most of the urbanization is set to take place in the lesser developed part of the world—by 2030, the towns and cities of the developing countries will make up 81% of the urban population (Kennedy 2008). Satisfactorily meeting the expected growth in energy demand is a major challenge for the international energy scenario especially when fossil fuels making up almost 80% of the global supplies are depleting.

A healthy and safe environment is important for human well-being, socio-economic prosperity of a society, and biodiversity at large. Global warming is argu-ably the biggest challenge faced by the mankind. In a survey, 50 Nobel laureates described climate change as the biggest threat facing mankind ahead of issues such as disease, nuclear war, and terrorism (Grove 2017). Climate change as a result of global warming is resulting into wide-ranging problems such as seasonal disorder, a pattern of intense and more frequent weather-related events such as floods, droughts, storms, heatwaves, wildfires, health problems, and financial loss. The United Nations (UN) warns that climate change can also lead to a global food crisis (Flavelle 2019). It has been reported that since the advent of the 20th century, natural disasters such as floods, storms, earthquakes, and bushfires have resulted into an estimated loss of nearly 8 million lives and more than US$7 trillion of economic loss (KIT 2016). Future projections suggest that by 2060 more than one billion people around the world might be living in areas at risk of devastating flooding due to climate change (McMillan et al. 2014). The majority of those affected will be from developing coun-tries with limited resources to mitigate the challenges and to rebuild their lives after extreme environmental events.

Adequate and affordable energy, healthy natural environment, and biodiversity are vital for addressing the socio-economic challenges including poverty, hunger, disease, and illiteracy. Poor and inadequate access to secure and affordable energy is hindering the progress of developing countries. Electricity, for example, is vital for providing basic social services such as education and health, water supply and purification, sanitation, and refrigeration of essential medicines. Electricity can also help support a wide range of income-generating opportunities. Although during the past 25 years, more than 1.5 billion people living in developing countries have been provided access to electricity, more than 800 million people still do not have access to it. Furthermore, more than 2.8 billion people rely on traditional biomass, including wood, agricultural residues, and dung, for cooking and heating. Statistics also suggest that more than 95% of the people without electricity live in developing regions, and four out of five live in rural areas of South Asia and sub-Saharan Africa. Climate change in terms of implications is also a concern mainly for developing countries. Though climate change is a global issue by its very nature and will affect all countries, the poorest will suffer the earliest and the most (Stern 2007). Ironically, the poor and developing countries, despite having a marginal contribution toward greenhouse gas emissions, are mainly at the suffering end when it comes to the consequences of climate change. Low-lying island nations are particularly facing serious challenges from the rising sea level. The grave ramifications of global warming and climate change call for an urgent paradigm shift in human activities and the use of natural resources.

With the growing world population and people's innate aspirations for an improved life, a central and collective global issue in the 21st century is to sustain socio-economic growth within the constraints of the Earth's limited natural resources, while at the same time preserving the environment. This target—sustainable development—can only be met by ensuring energy and environmental sustainability. The Sustainable Development Goals (SDGs) adopted by the UN in 2015 as part of its 2030 Agenda for Sustainable Development have also placed a strong emphasis on the energy and environmental sustainability. The SDGs focusing on the energy and environmental sustainability are highlighted in Table 1.1 (UN 2020).

The SDGs have replaced the Millennium Development Goals (MDGs); the latter were adopted by the UN in 2000 to address issues such as poverty, hunger, disease, illiteracy, and the environment. Of the eight MDGs, only one was directly relevant to the environment: the seventh MDG aimed to "Ensure Environmental Sustainability". The MDGs overlooked the energy challenges. The SDGs have taken the energy and environmental challenges seriously; 8 of the 17 SDGs are focused on energy and environmental sustainability. To achieve the SDGs, meaningful and integrated efforts are required on the part of all stakeholders including but not limited to the public and private sector, civil society, and citizens, to make the planet a better living place for the current and future generations.

1.2 SOUTH ASIA IN PERSPECTIVE

South Asia is the southern region of the Asian continent comprising the sub-Himalayan countries as shown in Figure 1.2. The total area of South Asia is around 5.2 million km². It is one of the most densely populated regions in the world, housing almost a quarter of the global population living in roughly 3% of the planet's total area. Traditionally, the region has been regarded to consist of seven countries: Bangladesh, Bhutan, India, Maldives, Nepal, Pakistan, and Sri Lanka. In 1985, these

TABLE 1.1

Overview of the Key SDGs Targeting Energy and Environmental Sustainability

Sustainable Development Goal	Description
SDG 6: Clean Water and Sanitation	Ensure availability and sustainable management of water and sanitation for all
SDG 7: Affordable and Clean Energy	Ensure access to affordable, reliable, sustainable and modern energy for all
SDG 11: Sustainable Cities and Communities	Make cities and human settlements inclusive, safe, resilient and sustainable
SDG 12: Responsible Consumption and Production	Ensure sustainable consumption and production patterns
SDG 13: Climate Action	Take urgent action to combat climate change and its impacts
SDG 14: Life Below Water	Conserve and sustainably use the oceans, seas and marine resources for sustainable development
SDG 15: Life on Land	Protect, restore and promote sustainable use of terrestrial ecosystems, sustainably manage forests, combat desertification, and halt and reverse land degradation and halt biodiversity loss

Source: UN 2020.

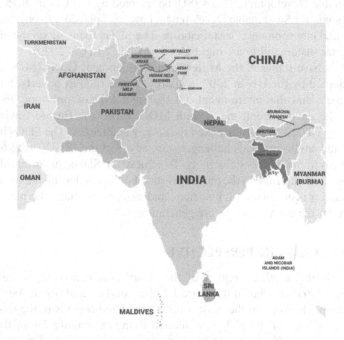

FIGURE 1.2   Map of South Asian Countries.

countries established a regional intergovernmental platform 'South Asian Association for Regional Cooperation' (SAARC) to promote economic cooperation and development in the region. In 2007, Afghanistan was formally included in the SAARC.

South Asia is a vibrant region with a great deal of diversity. The South Asian Countries (SACs) have commonalities as well as contrasts. There are religious, cultural, and historic commonalities. Most of the countries in the region, for example, gained independence from the British Empire after World War II. Socio-economic conditions in South Asian countries are largely in close proximity. India has the second largest population in the world after neighboring China, while Pakistan and Bangladesh are, respectively, fifth and eighth in terms of population. The contrasts are also quite obvious. On the one hand, the region has low-lying countries such as Maldives and Bangladesh, which are severely threatened by the rising sea level, and on the other hand, it has Nepal and Pakistan having most of the world's top 10 highest mountains including the two highest, Mount Everest and K2. In terms of population, it has Maldives and India with respective populations of almost half a million and over 1.380 billion as shown in Table 1.2. In terms of per capita gross domestic product (GDP), Afghanistan and Maldives have respective values of US$521 and US$10,331. In terms of area, Maldives and India have respective figures of 300 km^2 and almost 3.3 million km^2.

South Asia cherishes an enriched history, culture, geographic landscape, biodiversity, and natural resources. With Indus Valley, the region has several of the world's oldest civilizations. It is the birthplace of Hinduism and Buddhism and also is home to the largest Muslim population compared to any other part of the world. It has been an ancient trade route between the East and West. The vast population in the region faces wide-ranging socio-economic challenges, as all of the SACs fall under the typical categories of developing or underdeveloped countries. This was, however, not always the case. Before the British colonial era, a major part of the region known as the Indian Subcontinent—mainly comprising the

TABLE 1.2

Overview of SACs

Country	Population	Area (km²)	Population Below Poverty Line (%)	GDP/Capita (US$)
Afghanistan	38,928,346	652,230	54.5	521
Bangladesh	164,689,383	147,570	24.3	1,698
Bhutan	771,608	38,394	12.0	3,243
India	1,380,004,385	3,287,590	21.9	2,010
Maldives	540,544	300	15.0	10,331
Nepal	29,136,808	147,181	25.2	1,034
Pakistan	220,892,340	881,912	29.5	1,482
Sri Lanka	21,413,249	65,610	6.7	4,102

Data Sources: WM 2020; WB 2020; CIA 2020.

present-day Bangladesh, India, and Pakistan—was a byword for wealth. The Indian subcontinent has been historically known as a home for many commodities such as precious stones, spices, cotton, and textiles. This richness of the region paved the way for its poverty by attracting invaders and adventurers throughout the history from within Asia to as far as Europe. The latest episode in this respect, British Colonialism of the subcontinent, is widely acknowledged to have systematically drained the subcontinent of its resources (Tharoor 2017). The SACs have also to share the blame for some of their misfortunes, as they have traditionally suffered from disparities, with the low-caste "untouchables" condemned to dire poverty (Prasad 1999).

In the current global landscape, South Asia has obvious merits to draw strength from, while also facing issues of serious concern. On a positive note, the region has a large and thriving population base. With over 55% of the population under the age of 30 and only 9% beyond the working age of 65 years (LP 2020), the region has huge potential for development. It is regarded as an emerging region with significant improvements in recent years on the socio-economic and technological fronts as has been demonstrated by a healthy economic progress. The regional economy has benefited from investments in infrastructure, manufacturing, and agriculture (Seth 2020). According to the World Bank, between 2013 and 2016, SACs experienced an increase in growth from 6.2% to 7.5%, while the growth rates of other developing countries in general remained flat or even turned negative in some cases. Notwithstanding the last couple of years which have been tough for some of the SACs, especially India which is the largest economy in the region. The economic fallout of the COVID-19 pandemic, a universal crisis, is yet to be seen. South Asia offers the world a massive business and trade market and is being well appreciated for this. Conversely, the region still has a long way to go to accomplish its targets under the SDGs. There are ethnical biases and conflicts as well as religious extremism posing threats to peace and security in the region. South Asia faces serious energy and environmental problems and is suffering from wide-ranging geopolitical disputes and conflicts as discussed in the following section.

1.3   ENERGY AND ENVIRONMENTAL CHALLENGES

The South Asian countries are experiencing wide-ranging energy problems including, but not limited to, lack of access to refined fuels, weak grid reliability, unaffordable energy prices, and import dependency. These challenges are hampering the socio-economic prosperity of a large section of the region's population. Owing to factors such as a large and expanding population base, economic and infrastructure development, modernization, and urbanization, the region is expected to experience a growth in energy demand much faster compared to the global average. The provision of sufficient, refined, and affordable energy is key to sustainable development as also recognized in the SDGs. Large segments of South Asia's population have traditionally suffered from a lack of access to electricity and gas networks. In recent years, however, considerable progress has been made in terms of rural electrification making the national electricity-access numbers quite promising, as shown in Figure 1.3. The definition of electrification though is debatable. The definition

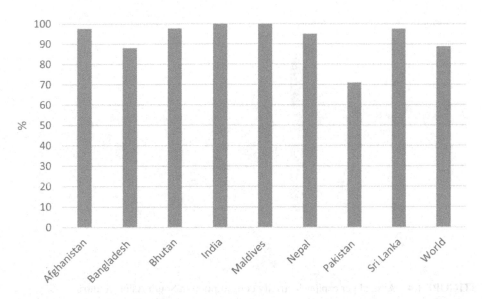

FIGURE 1.3 Electricity access in South Asian countries.

adopted by the Indian government, for example, is that 10% of a village should have electricity to have it declared electrified. Despite the Indian government's claim of having electrified every village in the country, up to 305 million people are reported to be lacking access to electricity (Heynen et al. 2019). In 2018 when India declared to have every village electrified, the average household electrification level in rural areas was around 80%, ranging from 47% to 100% across various states (TIE 2018). It is also noteworthy that electrification is not essentially an indicator of reliable and sufficient access to electricity. The reliability of the grid in many cases is feeble. Severe demand and supply issues are resulting in lengthy power outages and load-shedding even in urban centers and metropolitan cities. Issues with aging transmission and distribution (T&D) systems are also an acute problem in these countries. The fragility of the grid is also apparent in the form of power breakdowns and low voltages. These challenges have a major impact on South Asian countries' per capita electricity consumption as shown in Figure 1.4.

Lack of access to refined energy resources is also evident from the fact that over 1 billion people in South Asia meet their cooking and heating requirements through unrefined and polluting energy sources such as wood, animal dung, and crop waste as indicated in Table 1.3. In most of the SACs, biomass is the major resource for satisfying their cooking and heating requirements.

South Asia also suffers from acute environmental problems. It is one of the most vulnerable regions in the world in terms of the ramifications of global warming and climate change. Higher temperatures as a result of global warming are fast melting Himalayan glaciers, a lifeline for freshwater supplies which satisfy the drinking water and agricultural needs in the region. Rising sea level is posing numerous challenges to all of the SACs with coastal belts. Problems such as seawater intrusion, loss of wetland and mangroves, and displacement of human settlements are common

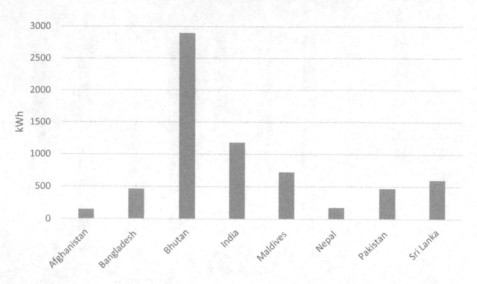

FIGURE 1.4   Annual per capita electricity consumption in South Asian countries.

in these countries. The situation with Maldives and Bangladesh is especially critical from the perspective of a shrinking land area. In the case of Bangladesh, for example, some scenarios suggest that by 2050, one-third of the country could be under water, making more than 70 million people homeless (Qudratullah and Asif 2020). Countries in the region are experiencing increased frequency and intensity of climate change-driven extreme weather events such as seasonal disorder, flooding, drought, and heatwaves. The Global Climate Risk Index 2020 regarded Pakistan, Bangladesh, and Nepal as the fifth, seventh, and ninth most affected countries in the world from extreme weather-related natural disasters over the period 1999–2018 (Eckstein et al. 2020). Climate change is also linked to water and food shortages. Water shortages and droughts are becoming a serious problem for these countries. In 2016, India reported having around 330 million people affected by severe drought, in many cases forcing people into internal migrations (BBC 2016).

South Asian countries are also facing environmental challenges directly associated with their use of energy. Air pollution, especially as a result of the use of polluting fuels and unchecked industrial and transport emissions, is also a major concern in the region. Pollution caused by the use of crude energy resources and inefficient technologies is a major environmental concern for these countries. The use of biomass fuels as shown in Table 1.3 leads to severe environmental and health issues. According to the World Health Organization (WHO), annually around 4 million people die prematurely from illnesses attributable to household air pollution from inefficient biomass-based cooking. South Asian countries are also experiencing exposure to dangerously high levels of particulate matter 2.5 (PM 2.5) as shown in Figure 1.5 (IQAir 2019). According to the 2019 World Air Quality Report, 21 of 30 of the world's cities with the worst level of air pollution are in India, including six of the top 10. Four of the top five countries in terms of PM 2.5 are in South

TABLE 1.3
Population Without Access to Clean Cooking Fuels

Country	Population (Million)	People relying on biomass (%)
Afghanistan	24.60	63
Bangladesh	125.40	76
Bhutan	0.18	23
India	704.00	51
Maldives	0.01	1
Nepal	20.60	71
Pakistan	123.8	56
Sri Lanka	14.80	69
Total	1,013.40	

Source: WB 2020

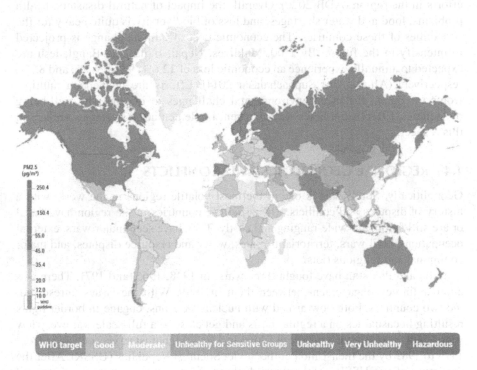

FIGURE 1.5   Global map of PM 2.5 exposure by country/region in 2019.

Asia—Bangladesh, Pakistan, Afghanistan, and India—at respective positions 1, 2, 4, and 5 in the world.

South Asian countries continually experience new challenges added to the environmental scenario. In recent years, India and Pakistan have started to face an annually recurring environmental problem in the form of smog. Mainly attributed to the agricultural, industrial, and transport activities, a thick layer of smog engulfs a large part of the two countries for several weeks, severely impacting the socio-economics and well-being of large segments of population. In November 2019, the Indian capital, Delhi, recorded air quality index of well over 900, far higher than the "severe-plus emergency" level of 500 (Stubley 2019), and PM2.5 levels of as high as 533 micrograms per cubic meter. The situation made Delhi's Chief Minister describe the city to have turned into a "gas chamber" (BBC 2019). Visibility is reduced to such an extent that highways, rail networks, and airports have to be closed even in cities such as Delhi and Lahore.

The environmental challenges are having serious economic implications in the region. Over half of the world's population categorized as "absolutely poor" live in South Asia, the livelihood of many of whom depends on climate-sensitive sectors such as agriculture, forestry, and fishing. Owing to its impacts on agriculture and the water cycle, climate change is affecting the food and water security. There is an inevitable knock-on effect on poverty alleviation and economic development efforts in the region (ADB 2020). Overall, the impact of natural disasters, health problems, food and water shortages, and loss of biodiversity is quite heavy for the economies of these countries. The economic cost of climate change is projected to intensify in the future. By 2100, Maldives, Nepal, India, and Bangladesh are expected to annually experience an economic loss of 12.6%, 9.9%, 9.4%, and 8.7%, respectively (Ahmed and Suphachalasai 2014). Efforts are needed on multiple fronts to combat the faced environmental challenges to improve the well-being of societies. Effective policies with stringent implementation strategies are key in this respect.

1.4 REGIONAL GEOPOLITICS AND CONFLICTS

Geopolitically, South Asia is one of the most volatile regions in the world with a history of disputes and conflicts. The issues the countries in the region have faced or are still facing are wide-ranging and costly. They have seen major wars, external occupations, civil wars, territorial disputes, water and resource disputes, and major communal and religious riots.

India and Pakistan have fought three wars, in 1948, 1965, and 1971. There was another limited engagement between them in 1999. With the issues unresolved, the two countries, both now armed with nuclear weapons, engage in border fights resulting in casualties on a regular basis and get close to a full-scale war every few years. India also fought a war with China in 1962. Afghanistan was occupied from 1979 to 1989 by the then-Union of the Soviet Socialist Republics (USSR). After the departure of the USSR, Afghanistan descended into a long sequence of internal chaos, civil wars, and post-9/11 US invasion. The country is yet to see peace and stability while millions of war-driven Afghan refugees are still living in neighboring countries. Pakistan experienced a civil war in 1971 combined with a war with India resulting in separation of its eastern part, "East Pakistan", which became an independent country, Bangladesh. Sri Lanka also has suffered from a 26-year-long civil war from 1983 to 2009.

1.4.1 TERRITORIAL DISPUTES

Territorial disputes are the biggest challenge undermining peace and security in South Asia. Most of the countries have boundary and territorial issues with their neighbors. India for example has disputes not only with three of its regional neighbors—Bangladesh, Nepal, and Pakistan—but also with China. India's territorial disputes with Nepal and China have reignited in recent years. In 2020, the row between India and Nepal has taken a major turn as the latter's parliament approved a new map of the country which includes some areas claimed by the former (Sharma 2020). The India–China dispute, along the nearly 3,500 km long Line of Actual Control (LAC) between the two countries, has also escalated in recent years after decades of relative peace. In 2017 troops from the two countries were engaged in a 73-day standoff over Doklam. The year 2020 has seen a major escalation in the situation involving another standoff, India's allegation of several kilometers of incursion inside its territory by China's People's Liberation Army, and a brawl between the border troops on 15 June, 2020 resulting into dozens of casualties. This being the first incident between the two nations in 45 years to result in loss of life, heavy military build-up by both nuclear-armed sides along the LAC amid COVID-19 is not a very promising sign for the region.

India and Pakistan have been engaged in a fierce dispute over the Jammu and Kashmir region since their independence from the British Raj in 1947. The Kashmir issue has been the point of contention between the two countries from their very beginning and they have fought three wars. Tension over the Kashmir issue hardly eases off and there have been multiple occasions when the two sides were on the brink of another war, the most recent being in 2019. The armies of the two countries equipped with nuclear weapons are almost in a continuous standoff. The unresolved Kashmir issue is widely acknowledged as a nuclear flashpoint. The two countries also blame each other for being involved in proxy wars and supporting separatist movements. The baggage of historic, religious, cultural, and geopolitical rivalries and the trust deficit between the two neighbors continue to spill over other areas and avenues. The latest addition to the catalog of geopolitical issues is the confrontation around India's objection to the Chinese investment and infrastructure development in Pakistan under the China Pakistan Economic Corridor (CPEC) initiative as part of China's "One Road One Belt" plan.

1.4.2 WATER DISPUTES

Another important challenge the region is facing is water disputes. Most of the SACs are water-stressed, and rivers are the lifelines for their agricultural and freshwater requirements. There are many disagreements between the regional countries on the share of transboundary river flows. Originating from the Himalayan basin, three major river basins are serving the region. These include Indus, Ganges, and Brahmaputra (Bala 2018). The Indus River flows westward toward the Arabian Sea, while the remaining rivers largely flow eastward toward the Bay of Bengal. Construction of dams and barrages in the upper riparian locations without consideration for the needs of the lower riparian areas is typically the main confrontational issue. Water is a precious commodity in agrarian societies and its importance is ever-increasing in South Asia due to a number of factors including growing water demands, shrinking water resources due to climate change and variations in the water cycle, agriculture-dependent socio-economics, weak water treaties, and mutual distrust. It is also noteworthy that some of the rivers in the region also carry religious and cultural sensitivities (Khan 2017). SACs have long-standing water disputes. While some of these countries have their internal disputes between states and provinces over the share of water from rivers and canals, there are also major issues internationally as highlighted in Table 1.4. On top of the changes in the water cycle and river flows, engineering interventions to build water storage dams and barrages are complicating the situation. Some of the notable disputes over rivers within the regional countries are between Bangladesh and India, India and Nepal, and India and Pakistan. There is also a dispute between India and Sri Lanka about the fishery in the Palk Bay.

The most critical of water disputes is the one between India and Pakistan. Issues on the distribution of water from the rivers flowing downstream from India to Pakistan started from the very beginning of the two countries. The disagreements over water rights were addressed through the Indus Waters Treaty of 1960, brokered by the World Bank, to fix and delimit the rights and obligations of both countries over the use of the waters of the Indus River System comprising of six rivers: Indus, Beas, Jhelum, Chenab, Ravi, and Sutlej. The treaty appears to be eroding as Pakistan perceives the recent upstream developments in India, especially the construction of dams, a violation of the Indus Water Treaty, and the two sides have been thus far unsuccessful in mutually resolving the issues. Given the importance of the Indus basin water for both countries, especially for Pakistan, which almost entirely relies on

TABLE 1.4
Overview of Geopolitical Issues Facing South Asia

Type of Issue	Involved Countries
Wars	India–China (1962)
	India–Pakistan (1948, 1965, 1971, 1999)
Foreign Invasions	Afghanistan's invasion by USSR
Civil Wars	Pakistan (1971)
	Sri Lanka (1983–2009)
	Afghanistan (1989–1996)
Ongoing Issues	
Territorial Disputes	Bangladesh–India
	India–China
	India–Nepal
	India–Pakistan
Water Disputes	Bangladesh–India
	India–Nepal
	India–Pakistan
	Indi–Sri Lanka

it for agricultural and freshwater requirements, and livelihood of over three-quarters of its population base, the water dispute is a serious issue. Pakistan feels aggrieved that India is taking undue advantage of its upper riparian position, something India denies. The current Indian government plans to review the Indus Water Treaty to use water as an instrument of foreign policy, labeling it as "water war" to pressure downstream Pakistan (TET 2018). The rhetoric by the Indian Prime Minister to block the flow of rivers to Pakistan as a punitive measure has gained momentum in the wake of the ongoing unrest in the Indian controlled part of Kashmir (Ghoshal and McKenzie 2019). Such a move is bound to have devastating implications for Pakistan which has declared to treat such an action as "an act of war". Given the historic baggage of distrust between the two nuclear-armed countries, which have fought wars over the unresolved issue of Kashmir, the water dispute becomes not only extremely complicated and critical for the two sides but also a matter of grave concern for the geopolitical stability of the entire region.

South Asian countries also experience religious and communal divides and systematic biases which sometimes spill over as international issues. There is a history of religious and communal riots, sometimes even flaring up tensions between the regional countries. SACs have largely fallen short of demonstrating pragmatism and maturity to resolve their outstanding issues with neighbors. In a prevailing environment of distrust and lack of cooperation, SAARC has become a virtually dysfunctional body. Countries in the region are annually spending over US$80 billion on defense and security apparatus, while hundreds of millions of their inhabitants live below the poverty line. With most of their socio-economic indicators being unenviable, the sustainable development goals especially around energy and environment look highly unachievable.

1.5 STRUCTURE OF THE BOOK

This book provides a holistic account of the energy and environmental outlook for South Asia. The scope and structure of the book are designed with the view to serve this objective. The book consists of 11 chapters. Chapter 1, Introduction, discusses the significance of the energy and environmental sustainability. It also presents a detailed geopolitical perspective of South Asia in terms of the wide-ranging issues undermining not only the stability and development in the region but also its energy and environmental outlook. The subsequent eight chapters are country-specific. The energy and environmental scenarios of each of the South Asian nations are discussed in separate chapters. These eight chapters, Chapters 2 to 9, are organized in the book in alphabetical order by country name: Afghanistan, Bangladesh, Bhutan, India, Maldives, Nepal, Pakistan, and Sri Lanka. All these chapters are quite focused within the scope of the book in providing comprehensive energy and environmental perspectives of the respective countries also taking into account key challenges and the prospective solutions. Chapter 10 provides a regional perspective on the energy and environmental scenes of South Asia. Chapter 11, Conclusions, summarizes the region's challenges and prospects in terms of a sustainable energy and environmental outlook. It discusses the importance of regional cooperation to improve energy and environmental security. It also reflects upon the new-normal scenario in the wake of the COVID-19 pandemic and the prospects for improvements on the energy and environmental fronts with the "build back better" approach.

REFERENCES

Ahmed M and Suphachalasai S (2014). Assessing the Cost of Climate Change and Adaptation in South Asia, Asian Development Bank, ISBN 978-92-9254-511-6

Asif M (2011). Energy Crisis in Pakistan: Origins, Challenges and Sustainable Solutions, Oxford University Press, Karachi

Bala A (2018). South Asian Water Conflicts, South Asia Journal, http://southasiajournal.net/south-asian-water-conflict/

BBC (2016). India drought: "330 million people affected", BBC, 20 April 2016

BBC (2019). Millions of masks distributed to students in "gas chamber" Delhi, BBC, 1 November 2019, https://www.bbc.com/news/world-asia-india-50258947

BP (2019). BP Energy Outlook: 2019 edition, British Petroleum

CIA (2020). The World Factbook, Central Intelligence Agency, https://www.cia.gov/library/publications/the-world-factbook/fields/221.html

EIA (2017). EIA projects 28% increase in world energy use by 2040, Today in Energy, Energy Information Administration, 14 September 2017, https://www.eia.gov/todayinenergy/detail.php?id=32912

EIA (2019), Today in Energy, EIA projects nearly 50% increase in world energy usage by 2050, led by growth in Asia, https://www.eia.gov/todayinenergy/detail.php?id=41433

EIA (2019), Today in Energy, EIA projects nearly 50% increase in world energy usage by 2050, led by growth in Asia, https://www.eia.gov/todayinenergy/detail.php?id=41433

Flavelle C (2019). Climate Change Threatens the World's Food Supply, United Nations Warns, New York Times, 8 August 2019

Eckstein D, Künzel V, Schäfer L, and Winges M (2020). Global Climate Risk Index 2020, German Watch

Ghoshal D and Mackenzie J (2019). India reiterates plan to stop sharing of excess water with Pakistan: minister, Reuters, 21 February 2019

Grove J (2017). Do great minds think alike? The THE/Lindau Nobel Laureates Survey, Times Higher Education, 31 August 2017

Heynen A, Lant P, Sridharan S, Smart S, and Greig C (2019). The role of private sector off-grid actors in addressing India's energy poverty: An analysis of selected exemplar firms delivering household energy, Energy and Building, 191, 95–103

IEA (2020). Energy security, International Energy Agency, https://www.iea.org/topics/energy-security

IQAir (2019). 2019 World Air quality Report, IQAir, https://www.iqair.com/world-most-polluted-cities

KIT (2016). Natural Disasters Since 1900: Over 8 Million Deaths and 7 Trillion US Dollars Damage; Press Release 058/2016; Karlsruhe Institute of Technology: Karlsruhe, Germany, 2016

Khan M (2017). Geopolitics of water in South Asia, Journal of Current Affairs, 1, Nos. 1&2: 66–86

Kennedy K (2008). Habitat for humanity and urban issues, The Forum, 15, 1

LP (2020). Population of South Asia, Live Population, https://www.livepopulation.com/country/south-asia.html

McMillan M, Shepherd A, Sundal A, Briggs K, Muir A, Ridout A, Hogg A, Wingham D (2014). Increased ice losses from Antarctica detected by CryoSat-2. American Geophysical Union. Geophys. Res. Lett, 41, 3899–3905.

Prasad H S (1999). Poverty in South Asia, Encyclopaedia Britannica, https://www.britannica
.com/topic/poverty-556003

Qudratullah H and Asif M (2020). Dynamics of Energy, Environment and Economy, Springer,
ISBN: 9783030435776

Seth S (2020). South Asian Countries: The New Face of Emerging Economies, Investopedia,
17 Mar 2020; https://www.investopedia.com/articles/investing/022316/south-asia-new-
face-emerging-economies.asp

Sharma G (2020). Nepal's parliament approves map including territory controlled by India, Reuters, 18 June 2020, https://www.reuters.com/article/
us-nepal-india-map-idUSKBN23P138

Stern N (2007). The Stern Review, The Economics of Climate Change, Cabinet Office, HM
Treasury, UK 2007

Stubley P (2019). Toxic smog forces flight cancellations, diversions and delays as worst air
pollution this year hits Delhi, Independent, 3 November 2019, https://www.indepen-
dent.co.uk/news/world/asia/toxic-smog-dehli-worst-air-pollution-year-flight-cancella-
tions-diversions-delays-a9183521.html

TIE (2018). Village electrification definition has lost relevance: Centre, 1 May 2018,
https://indianexpress.com/article/india/centre-clarifies-definition-of-electrification-
of-villages-5157644/

Tharoor S (2017). Inglorious Empire: What the British Did to India, 1st Edition, Hurst

TET (2018). Modi government lays groundwork for water war in battle with rival Pakistan, The
Economic Times, 12 July 2018, https://economictimes.indiatimes.com/news/defence/
modi-government-lays-groundwork-for-water-war-in-battle-with-rival-pakistan/
articleshow/54936280.cms

UN (2019). Department of Economic and Social Affairs, United Nations, https://www.un.org/
development/desa/en/news/population/2018-revision-of-world-urbanization-prospects
.html

UN (2020). Sustainable Development goals, United Nations, https://sustainabledevelopment
.un.org/?menu=1300

WB (2020). GDP per Capita, World Bank, https://data.worldbank.org/indicator/NY.GDP
.PCAP.CD

WM (2020). Worldometers, https://www.worldometers.info/world-population/population-by-
country/; https://www.worldometers.info/geography/largest-countries-in-the-world/

2 Afghanistan's Energy and Environmental Scenario

Najib Rahman Sabory[1], Mir Sayed Shah Danish[2], and Tomonobu Senjyu[3]

CONTENTS

2.1 Country Overview 17
2.2 Key Stakeholders of the Energy and Environment Sectors of Afghanistan 21
2.3 Afghanistan's Energy Outlook 22
2.4 Key laws and Regulatory Framework and Policies of the Energy and Environment Sectors of Afghanistan 24
    2.4.1 Environment 24
    2.4.2 Energy 29
2.5 Afghanistan's Environmental Status (Urban and Rural) 30
    2.5.1 Energy Sector GHG Emissions 33
    2.5.2 Industrial Processes GHG Emissions 33
    2.5.3 Agriculture Sector GHG Emissions 33
    2.5.4 Land-Use Change and Forestry-Related GHG Emissions 35
2.6 Conclusion 35
References 36

2.1 COUNTRY OVERVIEW

Afghanistan is a landlocked country connecting Central Asia and South Asia, covering slightly over 652,000 km^2 of area. It has borders with Turkmenistan, Uzbekistan, and Tajikistan to the north, China to the northeast, Pakistan to the east and south, and the Islamic Republic of Iran to the west. Its climate is continental arid and semi-arid with cold winters and hot summers. It is a mountainous country with an average elevation of 1100 meters from the sea level ranging from 250–8000 meters. Precipitation is mostly fallen as snow in winter and rain in spring (NCSA-NAPA, 2009; FAO, Aquastat, 2012). Figure 2.1 shows the average monthly temperature and precipitation of Afghanistan from 1901 to 2016.

According to the Afghanistan National Statistical and Information Authority (NSIA), the population of the country is nearly 31 million, with 22.6 million living in rural areas. The annual growth rate of the population is estimated to be 2.14% for

---

[1] PhD Student, University of the Ryukyus, Japan & Head of Energy Department, Engineering School Kabul University, Afghanistan.
[2] Assistant Professor, Energy Department, Engineering School Kabul University, Afghanistan.
[3] Professor, Electrical and Electronics Department, University of the Ryukyus, Japan.

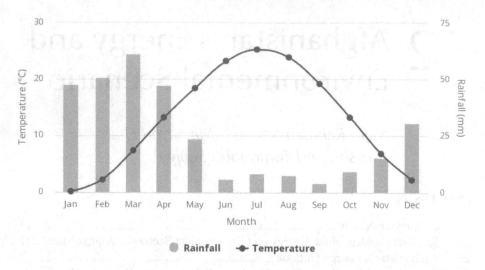

FIGURE 2.1   Afghanistan's average monthly temperature and precipitation from 1901–2016. (Source: WB, 2020.)

2015–2020. The gross domestic product (GDP), excluding poppy, was around US$20 billion with a per capita value of US$630, and a growth rate of 2.7% (NSIA, 2019). About one-seventh of the total land of Afghanistan is an agricultural area. Figure 2.2 shows the percentage of each type of agricultural areas.

Agriculture's contribution to the GDP, excluding poppy, was 18.6%, whereas industry and services contributed 24.1% and 52.6%, respectively, in 2019 (NSIA, 2019). Despite having a semi-arid climate, Afghanistan is still rich in water resources due to high mountains that provide almost 80% of the country's water (FAO, Aquastat,

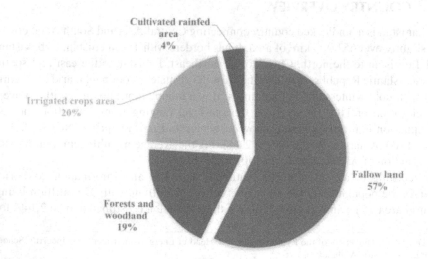

FIGURE 2.2   Agricultural lands classifications. (Source: NSIA, 2019.)

2012). Afghanistan's water resources are divided into five river basins namely Kabul, Helmand and Western, Harirod-Murghab, Northern, and Amu Darya. Figure 2.3 shows the river basin map of Afghanistan.

According to the UNDP Human Development Report for 2019, Afghanistan is ranked 170 among 189 countries with a Human Development Index (HDI) score of 0.496 (UNDP-HDR, 2019). Afghanistan's total available grid electricity generation capacity is 1550 MW, from which around three quarters is imported from Uzbekistan, Tajikistan, Turkmenistan, and Iran. The local production is mostly from hydropower plants that make up 44% of the total. Thermal power plants constitute 41% and diesel generators 15% of the remaining local generation capacity. Distributed all around the country, there are around 37 MW of off-grid small renewable energy generation plants, mostly from hydropower and solar energy (BSW, AREU, Eclareon, 2017). According to the Inter-Ministerial Commission for Energy (ICE) quarterly report, the electric energy consumption in 2015 was 4,847 MWh that give per capita consumption of 178 kWh that is amongst the lowest in the world (ICE, 2014). According to this report, electricity generation in 2016 from local sources has been from hydropower, diesel, and thermal with 96%, 3.5%, and 0.5% shares, respectively (ICE, 2014). According to the Ministry of Energy and Water, Afghanistan's renewable energy potential is estimated at around 300,000 MW consisting of 222 GW of solar energy, 66 GW of wind power, 23 GW of hydropower, and 4 GW of biomass (ITPower, 2017). Afghanistan's energy sector strategy strongly encourages the utilization of the domestic hydropower potential for its electricity needs and economic

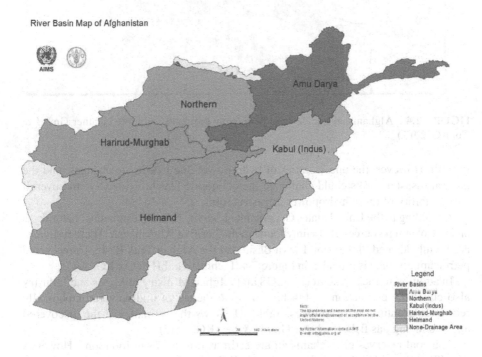

FIGURE 2.3   River basin map of Afghanistan. (Source: (Kamal & Raphy, 2004)

TABLE 2.1

Estimates of the Discovered and Producing Gas Fields

No.	Reserve Name	Remaining Gas Reserves (BCM)
1	Bashikurd	6.37
2	Yatimtaq	7.36
3	Jarquduq	9.77
4	Jangalikolon	13.38
5	Khoja Gogerdak	16.77
6	Juma	21.82
	Total	75.47

Source: Fichtner GmbH & Co. KG, 2013

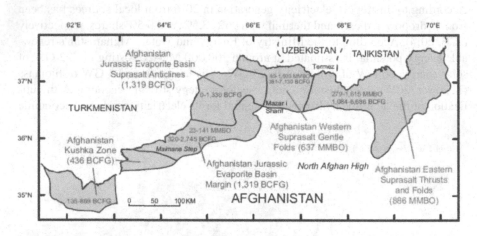

FIGURE 2.4 Afghanistan unrecovered petroleum estimates. (Source: Fichtner GmbH & Co. KG, 2013.)

growth. However, the uncertainties of hydropower due to climate changes, and the natural disasters risks could challenge the economic feasibility, safety, and overall sustainability of these hydropower infrastructures.

According to the United States Geographical Survey (USGS) estimation, natural gas and petroleum resources are mainly found in the north of Afghanistan. These resources are mainly located in the Amu Darya Basin and the Afghan Tajik Basin. Unrecovered petroleum estimate is provided in Figure 2.4 (Fichtner GmbH & Co. KG, 2013).

In addition to these resources, USGS and Afghan Ministry of Mines and Industry also claim the estimation of 444 billion cubic meters of undiscovered technically recoverable natural gas resources. Table 2.1 shows the estimates of the discovered and producing gas field (Fichtner GmbH & Co. KG, 2013).

The coal reserves of Afghanistan are estimated to be 73 million tons. However, the USGS states that the information is very little about Afghanistan's coal resources.

Coal is generally used for household cooking and heating in small scales, and cement production and power production, in some instances. Domestic oil production is insignificant, around 400 barrels of crude oil per day. Most of the petroleum consumed in the country is imported from neighboring countries.

2.2 KEY STAKEHOLDERS OF THE ENERGY AND ENVIRONMENT SECTORS OF AFGHANISTAN

Energy and environment sectors are interlinked and related to many other sectors. National Environmental Protection Authority (NEPA) and the Ministry of Energy and Water (MEW) are key organizations for the environment and energy of Afghanistan. In addition to NEPA and MEW, which is responsible for the overall policy development and governance of the country's environmental and energy issues, other key institutions that share the responsibility of addressing environmental, energy, and climate change and issues are illustrated in Figure 2.5. The full name of the organizations could be found in the footnote[4].

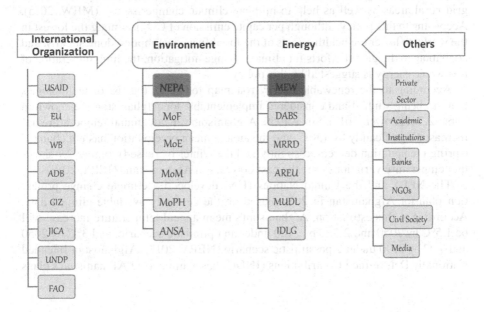

FIGURE 2.5 Key institutions that share the responsibility of addressing environmental, energy and climate change in Afghanistan.

---

[4] ADB: Asian Development Bank, ANSA: Afghanistan National Standards Authority, DABS: Da Afghanistan Brishna Sherkat, EU: European Union, FAO: Food and Agriculture Organization, GIZ: German International Cooperation, IDLG: Independent Directorate of Local Governments, JICA: Japan International Cooperation Agency, MEW: Ministry of Energy and Water, MoE: Ministry of Economy, MoF: Ministry of Finance, MoJ: Ministry of Justice, MoM: Ministry of Mines, MoPH: Ministry of Public Health, MRRD: Ministry of Rural Rehabilitation and Development, MUD: Ministry of Urban Development and Land, NEPA: National Environmental Protection Agency, UNDP: United Nations Development Program, USAID: United States Aid for International Development, WB: World Bank.

2.3 AFGHANISTAN'S ENERGY OUTLOOK

Afghanistan, a least developed country (UNCTAD-b, 2019) has one of the lowest per capita electricity consumption in the world, which is 178 kWh per year (BSW, AREU, Eclareon, 2017). It has a total of 1550 MW of generation capacity, from which two-thirds is imported from neighboring countries. Sixty-five percent of the urban population and 30% of the rural population has access to grid power. Afghanistan has good renewable energy potential, with 300 sunny days, 222 GW of solar energy potential, 66 GW of wind, and 23 GW of hydropower potential (MEW, 2019). Figure 2.6 describes the electricity status of Afghanistan.

According to Afghanistan's renewable energy policy, 95% of the electricity needed by 2032 shall be generated from renewable energy resources. It is estimated that the total need for electricity by 2032 would be nearly 6000 MW from which 4500–5000 MW is planned to be from renewables. This policy also emphasizes the technical, economic, and environmental sustainability of the energy sources. One of the goals of the Renewable Energy policy is to Help build a safe environment by reducing air pollution, safeguard human health and the environment, provide power to off-grid rural areas as well as help to mitigate climate change issues" (MEW, 2015). According to this policy, although per capita emission of CO_2 is among the lowest in the world, to lower the health effects of the fossil fuel consumption domestically, and contribute to the global efforts for climate change mitigation, the high integration of renewable energy is suggested by this policy.

According to the renewable energy roadmap for 2032, the list of technologies that are being studied and considered implementable for Afghanistan are shown in Table 2.2 (ITPower, 2017). Since 1950, Afghanistan's mean annual temperature has increased significantly by 1.8°C, and the annual mean precipitation has not changed (spring precipitation decreased slightly and the winter increased). Figure 2.7 shows the temperature variations over the past 60 years in Afghanistan (NEPA, 2017).

The NEPA and the United Nations (UN) developed a climate change projection plan for Afghanistan in 2016, based on the currently available climate data. According to this estimation, Afghanistan's mean annual temperature increase will be 1.5°C by 2050 and 2.5°C by 2100 under an optimistic scenario, and 3°C by 2050 and 7°C by 2100 under a pessimistic scenario (NEPA, 2017). Afghanistan Intended Nationally Determined Contributions (INDC) has summarized Afghanistan's plans

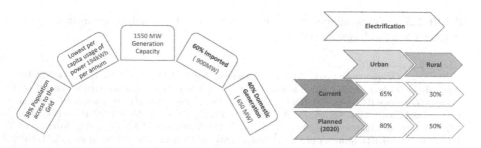

FIGURE 2.6 Energy status of Afghanistan.

TABLE 2.2

Studied and Applicable Renewable Energy Technologies for Afghanistan

Market	Technology	Type	Target Capacity (MW)
Utility scale	Solar PV	Solar power park (SPP)	590.0
	Solar PV	Solar independent power producer (IPP)	210.0
	Solar Thermal	Concentrating solar power (CSP)	110.0
	Solar PV	Floating PV	10.5
	Solar PV	Concentrating PV	6.0
	Wind	Wind IPP	600.0
	Large Hydro	Large hydro	2,000.0
	Biomass	Power generation from agri-residues	30.0
	Biogas - Electricity	Methanation of organic manure and agri-waste	2.0
	Waste to Energy	Bio-methanation of municipal solid waste	56.0
	Geothermal	Heat pump applications and geothermal energy	55.0
Mini-grid	Hybrid	Diesel + wind/solar/mini hydro	300.0
	MHP and SHP	MHP + SHP	420.0
Stand-alone	Solar PV	Rooftop with net-metering/FiT	420.0
	Solar PV	PV home systems, telecom towers and others	75.0
	Solar PV	Solar pumping	51.0
	Solar Thermal	Evacuative tube and flat plate - thermal energy	60.6
	Biogas	Thermal energy	4.1
	Improved Cookstoves	Thermal energy	2.6
Total			5,002.8

Source: ITPower, 2017.

and commitments for GHG emissions reduction. According to this plan, Afghanistan would need US$663 million per year for the deployment of the needed technology and capacity building efforts. This amount is broken down as US$100 million for the energy efficiency in buildings and transport sectors, US$188 million for the energy sector, US$74 million for the waste management, and another US$300 million for forest and rangeland, industry and mining, and agriculture and livestock sectors would be needed to reach the national targets. To meet the projected 5GW renewable energy generation capacity increase, the Afghanistan Renewable Energy Roadmap for 2032 proposes the renewable energy distribution by technology as shown in Table 2.3.

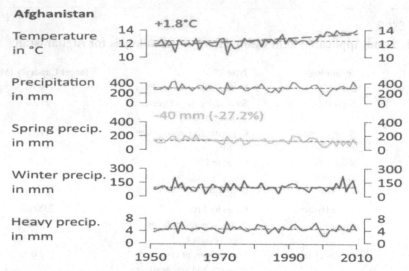

If trends are statistically significant (=0.5), they are plotted as dashed line and the magnitude is added.

FIGURE 2.7   Trends for temperature and annual precipitation in Afghanistan for the last 60 years. (Source: NEPA, 2017.)

2.4   KEY LAWS AND REGULATORY FRAMEWORK AND POLICIES OF THE ENERGY AND ENVIRONMENT SECTORS OF AFGHANISTAN

The laws, policies and regulations relevant to the energy and environmental sectors developed by different governmental entities are listed in Table 2.4:

2.4.1   ENVIRONMENT

The key legal documents related to the environment are described below.

*   **Environment Law (O.G. 912)**

    Article 15 of the Constitution of the Islamic Republic of Afghanistan obli-
    gates the state to adopt necessary measures for the protection of the living
    environment and forests (MOJ, 2004). The environment law was developed
    in 2007 for the realization of this state obligation. The main purpose of the
    Environment Law of Afghanistan is as follows: To improve livelihoods and
    protect the health of humans, flora and fauna; maintain ecological functions
    and evolutionary processes; secure needs of present and future generations;
    conserve natural and cultural heritage; and facilitate reconstruction and
    sustainable development of the national economy (MoJ-OG-912, 2007).
    According to this law, the National Environmental Protection Agency
    (NEPA) is set responsible for the implementation of this law, and the coor-
    dination, monitoring, conservation, and rehabilitation of the environment.
    According to this law, the rights and duty of the state are detailed as below:

TABLE 2.3
Distribution of Renewable Energy Breakdown by Technology for 2032

Market	Technology	Type	Target Capacity (MW)
Utility scale	Solar	Solar power park (SPP)	590
	Solar	Solar independent power producer (IPP)	210
	Solar	Concentrating solar power	110
	Solar	Floating PV	10.5
	Solar	Concentrating PV	6
	Wind	Wind IPPs	600
	Large hydro	Large hydro	2,000
	Biomass	Power generation from agri-residues	30
	Biogas	Organic and agri-waste	2
	Waste to energy	Municipal solid waste	56
	Geothermal	Heat pump applications and geothermal energy	55
Mini-grid	Hybrid	Diesel + wind/solar/mini hydro	300
	MHP and SHP	MHP + SHP	420
Stand-alone	Solar	Roof-top with net-metering/FiT	420
	Solar	PV home systems, telecom towers and others	75
	Solar	Solar pumping	51
	Solar	Evacuative tube and flat plate - thermal energy	60.6
	Biogas	Thermal energy	4.1
	Cook stoves	Thermal energy	2.60
Total			5,002.80

- Has sovereign rights over all biological and other natural resources in areas within the limits of its national jurisdiction.
- Is responsible for conserving the natural resources of the country and for sustainably using those resources, according to the provisions of the Environment Law.
- Has the duty, for the welfare of people, both present and future, to adopt and implement programmes aimed at maintaining and reestablishing essential ecological processes and natural resources, conserving and rehabilitating the environment, preventing and controlling pollution and ensuring that the benefits of the use of genetic resources are shared equitably.
- Has the duty to provide the public with information and opportunities to participate in making decisions affecting human health, the environment, and natural resources.

TABLE 2.4

Laws, Policies and Regulations Relevant to the Energy and Environmental Sectors Developed by Different Governmental Entities

Sector	No.	Name of the Document
	1	Environment Law (O.G. 912)
	2	Minerals and Hydrocarbons Law (O.G. 972)
	3	Mining Law (O.G. 1143)
	4	Afghanistan Green Urban Transport Strategy (AGUTS)
	5	National Environment Strategy
	6	National Housing Policy
	7	National Waste Management Policy
	8	National Environmental Action Plan (NEAP)
Environment	9	Clean Air Regulation (O.G. 991)
	10	Environmental Impact Assessment Regulations (O.G. 939)
	11	Regulation on Reduction and Prevention of Air Pollution (O.G. 991)
	12	Regulations on Controlling Materials Destructive to Ozone Layer (O.G. 894)
	13	National Adaptation Program of Action (NAPA)
	14	Intended Nationally Determined Contribution (INDC)
	15	Afghanistan Climate change strategy and Action Plan
	16	National Capacity Needs Self-Assessment (NCSA) for Global Environmental Management
	17	National Environmental Impact Assessment Policy
	1	National Energy Sector Strategy
	2	Power Sector Master Plan (PSMP)
	3	National Renewable Energy Policy (NREP)
	4	Afghanistan National Peace and Development Framework (2017)
	5	Regulation for Electricity Regulatory Authority of Afghanistan (2019)
Energy	6	Power Services Regulation Act (2015)
	7	Nuclear Energy Law (O.G. 1182)
	8	Afghanistan Energy Efficiency Code for Building (2015)
	9	National Energy Policy (Draft)
	10	Rural Renewable Energy Strategy
	11	Fuel Consumption Regulations (2010)
	12	Regulation on Fuel Consumption of Agricultural Machinery (O.G. 667)
	13	Renewable Energy Roadmap (Draft)

Rights and duties of the citizens are also described in the Environment Law as follows:

- Legally use natural resources under customary traditions and practices which encourage community-based sustainable natural resource management.
- Create and legally register civil society organizations that advocate the sustainable management of natural resources and conservation and rehabilitation of the environment.

- Have access to information held by the state, conditional to specific conditions.
- Participate in meetings, demonstrations, protests, marches, and referenda relating to the sustainable use of natural resources and conservation and rehabilitation of the environment according to this law.
- Address for consideration to the government and public bodies letters, complaints, declarations and proposals concerning sustainable use of natural resources and conservation and rehabilitation of the environment.
- Participate in environmental impact assessment processes.
- Refuse to undertake any work that would result in an imminent and serious threat to the environment or human health.
- Demand written justification regarding a particular decision.
- Appeal any decision made under this law in terms of Articles 19 and 77 of this law.
- Comply with this law and any other laws encouraging the sustainable use of natural resources and conservation and rehabilitation of the environment.

- **Hydrocarbons Law**
  The Hydrocarbons Law was enacted in 2009 according to Article 9 of the Constitution of Afghanistan to regulate the affairs related to determining state ownership and control over hydrocarbons, the preservation, utilization, and granting of concession rights, the execution of contracts, exploration activities, and the development and production of oil and gas. The objectives of this law are summarized as below (MoJ-OG-972, 2009; MoJ-OG-912, 2007):
  - To regulate the development and appropriate use of the mineral resources of Afghanistan;
  - To regulate and manage the reconnaissance, exploration and exploitation activities of mineral resources in Afghanistan;
  - The economic self-sustainability of Afghanistan through the development of its minerals sector;
  - To ensure that mineral resources are developed and managed according to the best international practices and experiences;
  - To secure optimal benefit from mineral extraction and processing;
  - The sustainable development of mineral resources, the prevention of waste, and the mitigation of negative environmental and social impacts;
  - To establish a suitable environment for national and international investment in the mining sector; and
  - To promote peace and security through the development of social and economic activities in the mining local communities.

- **Minerals Law**
  Afghanistan's Minerals Law was enacted based on Article 9 of the Constitution of Afghanistan in 2014. This law regulates all reconnaissance, exploration, developmental, and exploitation activities, and all types of mineral resources, mineral activities, and other ancillary activities in Afghanistan. Water and natural petroleum are excluded from this law. The

Ministry of Mines and Petroleum is the implementing agency of this law. The objectives of this law are:

- To regulate the development and appropriate use of the mineral resources of Afghanistan;
- To regulate and manage the reconnaissance, exploration and exploitation activities of mineral resources in Afghanistan;
- The economic self-sustainability of Afghanistan through the development of its minerals sector;
- To ensure that mineral resources are developed and managed according to the best international practices and experiences;
- To secure optimal benefit from mineral extraction and processing;
- The sustainable development of mineral resources, the prevention of waste, and the mitigation of negative environmental and social impacts;
- To establish a suitable environment for national and international investment in the mining sector; and
- To promote peace and security through the development of social and economic activities in the mining local communities.

- **National Capacity Needs Self-Assessment (NCSA) for Global Environmental Management**
  This document draws a vision to catalyze domestic and international assistance to build Afghanistan's capacities to effectively implement priority actions in the areas of biological diversity, climate change, and desertification, thereby contributing positively to sound global environmental management (NCSA-NAPA, 2009). The objectives of the NCSA are (NCSA-NAPA, 2009):
  - To identify priority issues for action within the thematic areas of biodiversity, climate change, and desertification;
  - To explore related cross-cutting and convention-specific capacity needs
  - To catalyze targeted and coordinated action and requests for future external funding and assistance; and
  - To link country action to the broader national environmental management and sustainable development framework.

- **National Adaptation Program of Action (NAPA) for Climate Change**
  The NAPA vision for Afghanistan is to increase awareness amongst all stakeholders of the effects of climate change and climate variability on their lives and to develop specific activities that build capacity to respond to current and future climate change threats (NCSA-NAPA, 2009). The objectives of NAPA are to (NCSA-NAPA, 2009):
  - Identify priority projects and activities that can help communities adapt to the adverse effects of climate change;
  - Seek synergies with existing MEAs and development activities with an emphasis on both mitigating and adapting to the adverse effects of climate change; and
  - Integrate climate change considerations into the national planning processes.

- **Afghanistan Green Urban Transport Strategy (AGUTS)**
  According to AGUTS, green urban transport in Afghanistan refers to a system of transport that is affordable, safe and secure, less harmful to the environment, accessible to all, and integrated and better managed. According to this strategy, the components of this system are comprised of vehicles, energy, infrastructure, roads (and maybe railways in the future), and terminals. The common modes of green urban transport in Afghanistan include bicycle, pedestrian, bus transport (especially using clean energy), light rail, non-motorized transport, and urban trucking.

  One of the strategies AGUTS focuses on is the reduction of emissions and greenhouse gases, to tackle the challenge of climate change, air quality, and health improvement which impact our high-level objective for protecting the environment and improving health. For the successful delivery of this strategy, AGUTS recommends working with the relevant government stakeholders including Afghanistan Environmental Protection Agency and Afghanistan National Standard Authority to enact and implement the emission allowable standards to improve the urban air, water, and land. It also suggests enabling cooperation among government agencies to tackle environmental challenges, promoting non-motorized modes of transport, focusing especially on the bicycle and pedestrian modes, and public transport to reduce car use, promoting and encouraging green technologies and phasing out highly polluting vehicles, and requiring and implementing environmental mitigation plans during the implementation of construction projects (Habibzai, 2014).

2.4.2 Energy

- **Power Sector Master Plan (PSMP)**
  PSMP was developed by the German consulting firm Fichtner GmbH with the funding support of the Asian Development Bank (ADB) in 2013. According to this plan, MEW is responsible for the implementation of this plan. This plan is made for 20 years (By 2032) for the Afghanistan power sector, with the main goal of increasing the electricity connection rate in an urban area to 100%, and 65% for the rural area households. It predicts the electrical energy demand by 2032 to be increased to 3500 MW with a higher scenario of 4500 MW. Moreover, the demand forecast, where generation and import supply options were analyzed and optimized. Additionally, a transmission network was designed to link supply and demand (Fichtner GmbH & Co. KG, 2013). In this plan, the environmental impact assessment of energy infrastructure-related projects, such as emissions, noise, water pollution, loss of fauna and flora habitats, and landscape disturbance, is suggested to be conducted when all design information and requirements are available.
- **Electricity Service Regulation Law**
  This law was enacted in 2015, the first law for the regulation of the electricity services in Afghanistan. The main objective of this law is the provision of

the needed electric energy from local resources as well as imported power, quality improvement and expansion of the power services, economic development of the country, fair tariff design for consumers, non-discriminatory access to electric energy, and regulation of the electricity market in the country. MEW is in charge of the implementation of this plan.

• **National Renewable Energy Policy (NREP)**
MEW is in charge of developing and implementing the NREP. The main purpose of this policy is to set a framework for the development of renewable energy in Afghanistan urban and rural areas. It targets deploying 4500–5000 MW of the power energy by 2032 from renewable energy resources that comprise 95% of the total electric energy demand by then. The scope of the policy covers the deployment of environmentally sustainable and techno-economically energy resources and technologies, all renewable energy resources and technologies that can be deployed in a techno-economically and environmentally sustainable manner in Afghanistan.

• **Afghanistan Energy Efficiency Building Code (AEEBC)**
AEEBC was developed by Afghanistan National Standards Authority in 2015. The main purpose of AEEBC is the promotion of energy-saving in building, lighting systems, building envelope, heating and cooling, and passenger conveyors. The harnessing of solar energy for buildings is also highly suggested by this code.

• **Afghanistan National Peace and Development Framework (ANPDF)**
ANPDF is the government's five-year (2017-2021) strategic framework for reaching stability and self-reliance. The ANPDF presents the country's immediate and long-term development plans by providing high-level guidance to the government and other stakeholders. Also, the ANPDF highlights Afghanistan's key reforms, outlines priority investments needed to achieve development goals in these critical areas, and sets the economic, political, and security context for sustainable development, focusing on agriculture, extractive industries, and trade. The ANPDF recognizes climate change as a serious threat to Afghanistan that needs to be addressed, particularly in the areas of agriculture production, increased risk of natural hazards arising from changing temperature and precipitation patterns, and renewable energy development to reduce greenhouse gas (GHG) emissions. Moreover, then ANPDF recommends increasing regional collaboration to mitigate the impacts of climate change and increase climate change adaptation across the trans-Himalayan region.

2.5   AFGHANISTAN'S ENVIRONMENTAL
STATUS (URBAN AND RURAL)

It is globally agreed that one of the main causes of climate changes that encompass several environmental issues is the excessive use of fossil fuel. The provision of sustainable energy is one of the key factors for a livable and safe environment and will largely help toward climate change mitigation. Climate change is defined by the United Nations Framework Convention of Climate Change (UNFCCC) as:

"a change of climate which is attributed directly or indirectly to human activity that alters the composition of the global atmosphere and which is in addition to natural climate variability observed over comparable periods." (UNFCCC, 1992). The earth's surface average temperature has increased by 0.74°C since the Industrial Revolution. It is predicted to increase by another 1.4°C to 5.8°C over the period 1990 to 2100 (IPCC, 2001), with the principal reason of a century and a half of consuming ever-greater quantities of oil, coal, the cutting of forests, and the practice of certain farming methods. These activities have augmented the amount of GHGs in the atmosphere (NCSA-NAPA, 2009). Afghanistan signed the UNFCCC on 12 June, 1992 as a Non-Annex I Party to the Convention but has not yet agreed to the Kyoto Protocol, therefore is currently unable to become a Clean Development Mechanism (CDM) host country. The ultimate objective of the UNFCCC that the Conference of the Parties (COP) has adopted is "to achieve, in accordance with the relevant provisions of the Convention, stabilization of greenhouse gas concentrations in the atmosphere at a level that would prevent dangerous anthropogenic interference with the climate system. Such a level should be achieved within a time-frame sufficient to allow ecosystems to adapt naturally to climate change, to ensure that food production is not threatened and to enable economic development to proceed in a sustainable manner" (NCSA-NAPA, 2009).

Afghanistan has a number of environmental problems such as a decrease in the water table, wetlands, and deforestation. It has been very severe in urban areas due to the unregulated and excessive usage of underground water. Continuation of this trend will create a potable water crisis in big cities such as Kabul, the capital of Afghanistan (FAO, Aquastat, 2012).

Other key environmental problems in Afghanistan's big cities include issues such as air, water, and soil pollution, and solid waste management. Urban areas are also the main contributors to climate change in this country. Although cities cover barely 2% of Afghanistan's land area, they consume around 70% of the country's energy and produce about half of its CO_2 emissions, as well as significant amounts of other GHG emissions. The main sources of these gases are energy generation, vehicles and transportation, the brick kiln industry, and biomass use (NEPA, 2017).

Main environmental pollutants in major urban areas in Afghanistan are contributed by the energy sector. The low access rate of urban population to reliable and sufficient electricity, worn and non-standard transportation system, use of fossil fuel and burnable municipal solid waste (plastic, used motor oil, cloth, and vehicles used tyres) for heating and cooking, lack of standards and regulations for buildings energy use, the low economy of the majority of the urban population, city geomorphology, low technical capacity in the environmental sector, and poor public awareness are key causes of urban pollution (Sabory, 2018). It is considered that environmental pollution is deadlier than war victims. State of the Global Air claims 26,000 deaths due to air pollution in 2017, while 3,484 civilians were killed in the war in the same year (Faiez, 2019). Figures 2.8 and 2.9 provide a glance at the environmental problems in Afghanistan and its capital, Kabul.

Afghanistan's GHG emissions composition in 2013 is detailed in Table 2.5.

The summary of the GHG emissions by sector is detailed in Table 2.6.

FIGURE 2.8   Kabul city air pollution from an altitude. (Source: Sabory, 2018.)

FIGURE 2.9   Unclean use of fossil fuel. (Source: Faiez, 2019.)

TABLE 2.5
Afghanistan GHG Emissions Composition in 2013 (Source: NEPA, 2017.)

Greenhouse Gas	Emissions (Gg)	Percentage of total	CO$_2$ Equivalent
CO$_2$	20,395	33.9	33.9
CH$_4$	519	31	18,684
N$_2$O	71	35.1	21158
Sector	Percentage		
Agriculture	64.3%		
Land-Use Change and Forestry	18.8%		
Energy	16.2%		
Industrial Processes	0.3%		
Waste	0.3%		

TABLE 2.6
GHG Emissions for Six Sectors in Afghanistan for 2013 (All Figures in Gg)
(Source: NEPA, 2017.)

GHG Source	CO$_2$ Emissions	N$_2$O	NO$_x$	CO	NMVOC	CO$_{e2}$
Total Emissions	20,395	73	70	541	45	60,237
Energy	9,639	0	61	235	45	9,747
Industrial Processes	210	0	0	0	0	210
Solvent and Other Product Use	NE	NE	NE	NE	NE	NE
Agriculture		72	4	110		38,762
Land-Use Change and Forestry	10,546	0	6	197		11,338
Waste		0				180

2.5.1 ENERGY SECTOR GHG EMISSIONS

In 2013, the energy sector of Afghanistan produced 87% of Afghanistan's NOx emissions (61 Gg), 43% of CO emissions (235 Gg), and almost 100% of non-methane volatile organic compounds (NMVOCs) (45 Gg). Emissions were estimated from fuel combustion (excluding biomass), with the transport sub-sector comprising 51.1 percent (4,924 Gg CO$_2$) of the total and the energy industries sub-sector comprising the remaining 48.9% (4,715 Gg CO$_2$). Emissions from the transport sub-sector constituted 72% of the NOx (44 Gg) and almost 100% of both CO (234 Gg) and NMVOC (45 Gg) for the energy sector as a whole. A summary of the GHG emissions from the energy sector is provided in Table 2.7.

2.5.2 INDUSTRIAL PROCESSES GHG EMISSIONS

Industries were collectively responsible for approximately 0.3% of total GHG emissions (210 Gg CO$_2$) for the country. Almost half of this came from lime production (100 Gg CO$_2$), followed by the production of cement (43 Gg CO$_2$), and iron/steel (38 Gg CO$_2$). Table 2.8 provides a summary of the GHG emissions from the industrial processes sector.

2.5.3 AGRICULTURE SECTOR GHG EMISSIONS

The agriculture sector (including both crops and livestock) accounted for approximately 94% of Afghanistan's overall CH$_4$ emissions (489 Gg or 17,604 Gg CO$_2$e). Given the high global warming potential of CH$_4$, this means that this sector is also the largest contributor (38,762 Gg CO$_{2e}$ or 64.3%) of overall GHGs emissions in the country. This primarily originated from enteric fermentation from livestock (414 Gg or 84%), with lesser amounts emitted in manure management (40 Gg or 8%) and cultivation of rice (29 Gg or 6%). Emissions of N$_2$O from agricultural soils constituted 72 Gg, which made up almost 100% of the emission of this gas in Afghanistan. The GHG emissions from the agriculture sector are summarised in Table 2.9.

TABLE 2.7
GHG Emissions for the Energy Sector in Afghanistan for 2013 (All Figures in Gg)

GHG Source	CO$_2$ Emissions	CH$_4$	N$_2$O	NO$_x$	CO	NMVOC
Energy Sector Total	9,639	3	<1	61	235	45
A. Fuel Combustion	9,639	1	<1	61	235	45
1. Energy Industries	4,715	<1	<1	18	1	<1
2. Manufacturing Industries and Construction	NE	NE	NE	NE	NE	NE
3. Transport	4,924	<1	<1	44	234	45
4. Other Sectors	NE	NE	NE	NE	NE	NE
B. Fugitive Emissions from Fuels	0	2		0	0	0
1. Solid Fuels		2		0	0	0
2. Oil and Natural Gas		NE	NE	NE	NE	NE

Source: NEPA, 2017.

TABLE 2.8
GHG Emissions for the Industrial Processes Sector in Afghanistan for 2013 (All Figures in Gg)

GHG Source	CO$_2$ Emissions	CH$_4$	N$_2$O	NO$_x$	CO	NMVOC
Industrial Processes Sector Total	210	0	0	0	<0.01	<0.01
A. Mineral Products	144					
B. Chemical Industry	<1	0	0	0	<0.01	<0.01
C. Metal Production	66					
D. Other Production	<1					
E. Production of Halocarbons and Sulphur Hexafluoride	NE	NE	NE	NE	NE	NE
F. Consumption of Halocarbons and Sulphur Hexafluoride	NE	NE	NE	NE	NE	NE
1. Solid Fuels		2		0	0	0
2. Oil and Natural Gas		NE	NE	NE	NE	NE

Source: NEPA, 2017.

TABLE 2.9
GHG Emissions for the Agriculture Sector in Afghanistan for 2013
(All Figures in Gg)

GHG Source	CH_4	NO_2	NO_x	CO
Agriculture Sector Total	489	72	4	110
A. Enteric Fermentation	414			
B. Manure Management	40	0	0	<0.01
C. Rice Cultivation	29			
D. Agricultural Soils		72		
E. Prescribed Burning of Savannas	0	0	0	0
F. Field Burning of Agricultural Residues	5	0	4	110

Source: NEPA, 2017.

2.5.4 LAND-USE CHANGE AND FORESTRY-RELATED GHG EMISSIONS

The land-use change and forestry sector forms the largest proportion (51.7%) of Afghanistan's CO_2 emissions at 10,546 Gg CO_2. It also contributes to the second largest proportion of the overall emissions of CO (197 Gg or 36%). Changes in woody biomass comprise 4,034 Gg (38%) of CO_2 emissions in this sector, while the conversion of forest and grasslands comprises the bulk of the emissions of all GHGs by this sector. See Table 2.10 for a summary of GHG emissions from the land-use change and forestry sector.

2.6 CONCLUSION

Energy and environment are interlinked topics, which need to be discussed jointly. These topics are widely linked to many other sectors such as economics, climate change, health, agriculture, water, security, and many more. To address environmental challenges, energy is the first topic to look at. In Afghanistan, many health issues, economic problems, water, air pollution, and local climate are somehow related to these major topics. The urban area is related and affected the most by energy and environmental issues. From the available data, it is clear that the agriculture sector is the main CO_2 contributor, and energy follows it. This is because Afghanistan is not an industrialized country. Growth in domestic production will accordingly affect energy consumption. At the same time, the environmental concern will equally increase as the production and industrial activities increase. The good news is that the government vision for 2032 is the production of 95% of the needed energy from the local renewable resources. However, achieving this goal would require conservative planning for the implementation of government's vision, close coordination with financial entities, the private sector, and human resource development entities. This also requires a strong commitment of the government and donor agencies, as Afghanistan is and will be relying or external financial support for a number of years. It is obvious that there is not a responsive legal framework in place for the energy and environmental

TABLE 2.10

GHG Emissions for the Land-Use Change and Forestry Sector in Afghanistan for 2013 (All Figures in Gg)

GHG Source	CO_2 Emissions	CH_4	NO_2	NO_x	CO
Land-Use Change and Forestry Sector	10,546	22	0	6	197
A. Changes in Forest and Other Woody Biomass Stocks	4,034	0	0	0	0
B. Forest and Grassland Conversion	6,512	22	0	6	197
C. Abandonment of Managed Lands	NE	NE	NE	NE	NE
D. CO_2 Emissions and Removals from Soils	NE	NE	NE	NE	NE

Source: NEPA, 2017.

issues in Afghanistan. Policymakers need to create an enabling environment where all sectors linked to energy and environment could work together and have required coordination to timely and efficiently work for the relevant concerns.

REFERENCES

BSW, AREU, Eclareon. (2017). *Enabling PV Afghanistan.* Kabul: GIZ.

Faiez, Rahim. (2019, Nov. 13). *AP News.* Retrieved from The Associated Press: https://apnews.com/1e566a9f6cd647d2998ec9c2582a6176

FAO, Aquastat. (2012). *Country Profile – Afghanistan.* Rome-Italy: Food and Agriculture Organization.

Fichtner GmbH & Co. KG. (2013). *Power Sector Master Plan.* Stuttgart, Germany: Asian Development Bank.

Habibzai, Abdullah. (2014). *Afghanistan Green Urban Transportation Strategy.* Kabul: Ministry of Transportation.

ICE. (2014). *Quarterly Energy Sector Status, Summary Report.* Kabul: MoE (Ministry of Economy of the Islamic Republic of Afghanistan).

IPCC. (2001). *Climate Change 2001: Impacts, Adaptation and Vulnerability.* Cambridge, UK: Intergovernmental Panel on Climate Change (IPCC).

ITPower. (2017). *Renewable Energy Roadmap for Afghanistan.* Kabul: ADB-IT Power Consulting Private Limited.

MEW. (2015). *Afghanistan Renewable Energy Policy.* Kabul, Afghanistan: Ministry of Energy and Water (MEW).

MEW. (2019, 11 01). *Ministry of Energy and Water.* Retrieved from www.mew.gov.af

MOJ. (2004, January 26). *The Constitution of the Islamic Republic of Afghanistan.* Kabul, Afghanistan: Ministry of Justice.

MoJ-OG-912. (2007, January 25). *Environment Law.* Kabul, Afghanistan: Ministry of Justice of the Islamic Republic of Afghanistan.

MoJ-OG-972. (2009). *Hydrocarbons Law.* Kabul, Afghanistan: Ministry of Justice of the Islamic Republic of Afghanistan.

NCSA-NAPA. (2009). *Afghanistan National Capacity Needs Self-Assessment for Global Environmental Management (NCSA) and National Adaptation Programme of Action for Climate Change (NAPA) Joint Report.* Nairobi, Kenya: UNEP.

NEPA. (2017). *Second National Communication Under the Climate Framework Convention on Climate Change (UNFCCC).* Kabul, Afghanistan: National Environmental Protection Agency.

NSIA. (2019). *Afghanistan Statistical Year Book 2018-2019.* Kabul: Afghanistan National Statistical and Information Authority (NSIA)..

Sabory, Najib Rahman. (2018, March 03). *8am Daily.* Retrieved from 8am Daily: https://8am .af/kabuls-pollution-disaster-or-opportunity/

UNCTAD-b. (2019). *The Least Developed Countries Report.* New York: United Nations.

UNDP-HDR. (2019). *Human Development Report.* New York: UNDP.

UNFCCC. (1992). *United Nation Framework Convention for Climate Change.* UN.

WB. (2020, March 03). *Climate Change Knowledge Portal.* Retrieved from World Bank Group (WBG): https://climateknowledgeportal.worldbank.org/country/afghanistan

3 An Overview of Energy Scenario in Bangladesh: Current Status, Potentials, Challenges and Future Directions

Imran Khan, Sujan Chowdhury, and Zobaidul Kabir

CONTENTS

3.1 Introduction 40
3.2 Institutional Arrangements for Energy Production 41
3.3 Renewable and Non-renewable Energy Sources in Bangladesh...................42
    3.3.1 Non-Renewable Sources 43
        3.3.1.1 Natural Gas 43
        3.3.1.2 Coal..44
        3.3.1.3 Petroleum 45
    3.3.2 Renewable Energy 46
3.4 Energy status in Bangladesh..47
3.5 Electricity scenario in Bangladesh 48
    3.5.1 Challenges and Future Directions 51
    3.5.2 Present Scenario 54
        3.5.2.1 Investment...56
        3.5.2.2 Technology Transfer and Human Resources
                 Development 56
        3.5.2.3 Affordable Fuel Mix 56
        3.5.2.4 Energy Efficiency and Conservation 56
        3.5.2.5 Fuel Utilization Plan 57
        3.5.2.6 Governance 57
    3.5.3 Energy and Environment..57
3.6 Policy Recommendations 58
3.7 Conclusions...60
References..60

3.1   INTRODUCTION

Bangladesh is one of the most densely populated and lowest energy-consuming nations in the world. Energy has a crucial role in improving living standards, which include education, healthcare, and other services, through which local, regional, as well as global development can be confirmed. In today's world, the development of any country is inexorably linked with energy production, supply, as well as consumption. In Bangladesh, about 70% of its energy demand is met from natural gas. Among other fuels, oil, coal, biomass, etc. are vital. There is reserve of coal in Bangladesh, but coal is less produced as well as less used there. On the other hand, the natural gas reserve is not substantial, but its production and consumption are the highest among the available resources. In addition to these, energy demand is being met through imported oil and liquefied petroleum gas (LPG). Moreover, the government has started importing liquefied natural gas (LNG) to meet increasing gas demand. Biomass is being used as the lion's share of energy. The energy demand is also being met by importing electricity from India. The use of renewable energy instead of gas, coal and oil has started throughout the world and is essential for sustainable development and protecting the environment by preventing carbon emissions. Several countries in the world, such as New Zealand, Sweden, Germany, China and USA, are currently using renewable energy as a significant part of their energy demand. For instance, New Zealand's electricity sector is dominated by hydro (Khan, Jack, and Stephenson 2018). Bangladesh is also using renewable energy, but it is not sufficient to meet the national sustainable development goals, see (Khan 2019b) and (Khan 2020c). The government has taken various steps to increase the use of renewable energy in the future, including solar home systems (SHSs) and solar irrigation systems. In addition, a nuclear power generation project has also been initiated by the government.

Bangladesh has several serious problems relating to financial growth due to its small and insufficient power supply. Out of 161 million people in Bangladesh, almost 63% live in rural areas (World Bank 2018). Therefore, Bangladesh focused on the utilization of renewable energy in rural areas in 1996, with its actualization occurring in 2002. In cooperation with the United Nations Development Programme, the "Renewable Energy and Energy Efficiency Programme" developed new energy policy (NEP) in 2006 and this was ratified in December 2009.

The government has shown commitment in creating an investment-friendly ecosystem by providing major financial incentives for independent power producers to overcome the anticipated decrease in the production volume of natural gas. In meeting the demand-supply deficit, it is expected that there will be a growing popularity in mobile financial services, such that it could attract more private investment toward installing new power plants on a build-own-operate (BOO) basis. The Rural Electrification Board (REB) has initiated a number of projects to cover all the villages across Bangladesh. Specifically, two separate projects and 2.7 million households will be connected to the national electric grid by 2021[1].

---

[1]   https://energypedia.info/wiki/Bangladesh_Energy_Situation (accessed on 1-Jan-2020).

The REB is currently implementing connection for roughly 5.8 million consumers and, consequently, there will be electricity coverage in every rural area of Bangladesh. In 2013, only an estimated 59.6% of the Bangladesh population was connected to the electricity grid and the maximal delivered electricity was 6,675 MW from 10,213 MW installed capacity (including public, private and import). In the 2018–2019 financial year, the total number of consumers, particularly the residences that got access to electricity was 213,078 (BPDB 2019). About 12% of the households have basic access to electricity through off-grid connections, such as solar home systems (Khan 2019b). The government estimates that more than 90% of the people now have access to electricity. In the 2018–2019 financial year, the total net electricity generation was 70533 GWh, which is an increase of about 12.53% from 62678 GWh in 2017–2018 (BPDB 2019). In the long-term, the government plans to connect 98% of households, mainly through grid extension, by 2021.

The aim of this chapter is to draw a picture of the current energy scenario in Bangladesh. The rest of the chapter is organized as follows: Section 3.2 discusses the institutional arrangements for energy production in Bangladesh; Section 3.3 highlights the renewable and non-renewable energy sources in Bangladesh; Section 3.4 focuses on the energy status of Bangladesh; Section 3.5 explains the present electricity scenarios in Bangladesh; Section 3.6 proposes possible policy implications for Bangladesh; and the final section, Section 3.7, concludes the chapter.

3.2 INSTITUTIONAL ARRANGEMENTS FOR ENERGY PRODUCTION

Proper pricing is a prerequisite for an uninterrupted supply of any quality product, especially the energy sector. However, living expenses are increasing due to higher energy prices, and this is crucial for any low-developed country such as Bangladesh. This is because the energy sector is subsidised by the government to reduce production costs. In addition, subsidies are also provided by electricity tariffs so that consumers can afford electricity; this has played a critical role in development of the Bangladesh economy (S. Islam and Khan 2017). However, the Bangladesh Power Development Board (BPDB) has recently taken different steps to purchase electricity from an independent power supplier to meet the growing demand for electricity. The total subsidy received from the government for power and energy sector in the financial year 2009 to 2020 is shown in Figure 3.1.

Energy policies play an essential role in the overall development of a country. Bangladesh is not an exception to this trend. Renewable energy harnessing, fuel diversification, energy efficiency and conservation, private sector participation in electricity generation, tariff rationalization, and regional cooperation regarding inter-country power trading are some of the strategic policies in the energy sector (Islam and Khan 2017). To encourage private sector participation, the Bangladesh government initiated the "Private Sector Power Generation Policy of Bangladesh". This policy will help attract more private investments, consequently guaranteeing competitive prices for power generation projects. BPDB has encouraged the use of

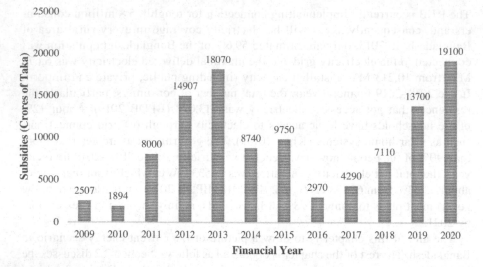

FIGURE 3.1 Government subsidy to power and energy from financial year 2009 to 2020. (Data source: Ministry of Finance[2]).

rooftop solar projects and connecting such systems to the national grid (Islam and Khan 2017). To reduce fuel imports for power generation, the government of Bangladesh initiated an energy efficiency and master plan, which might save 2.3 trillion BDT between 2015 and 2030 (EECMP 2015). In 2015, Bangladesh imported about 4% of its electricity from India through a cross-border trade policy; this was increased to 9% in 2019, see (Khan 2019a) and (BPDB 2019). The government has also opened a window to the importation of unutilized hydropower resources from Bhutan and Nepal. However, Bangladesh has one of the lowest per capita energy consumptions for the same per capita GDP of developed countries (EECMP 2015). Therefore, if the energy supply can be ensured, Bangladesh will be able to achieve higher economic growth such that it will become a developed nation by 2041. In addition, the Bangladesh government has to confront socio-economic and environmental impacts to control the subsidy and the generation of electricity.

3.3 RENEWABLE AND NON-RENEWABLE ENERGY SOURCES IN BANGLADESH

In Bangladesh, the estimated final consumption of total energy is around 47 Mtoe. On average, energy consumption increases about 6% per year. The per capita consumption of primary energy in Bangladesh was 0.22 toe (tonne of oil equivalent)

---

[2] https://www.thedailystar.net/bangladesh-national-budget-2019-20/gas-subsidy-double-1756864 (accessed on 29-Jan-2020)

in 2014, and this has increased since then (MPEMR 2018). For example, the per capita generation of electricity was 425.92 kWh and consumption was 374.62 kWh in the 2018–2019 financial year (BPDB 2019). According to the World Bank, the per capita consumption of energy in Bangladesh is lower than that of South Asian neighboring countries. The estimated primary energy consumption was raised to 47 Mtoe in 2018; this is almost double from 26.7 Mtoe in 2013 (Halder et al. 2015).

The investment from generating electricity over the last decade has increased. In Bangladesh, native commercial energy sources, such as natural gas, oil, imported LNG, LPG and electricity from India, hydroelectricity and coal, are the primary contributors to energy consumption. Of the total energy, biomass accounts for about 29% of the primary energy and the remaining 71% is met by commercial energy. In addition, Bangladesh imports about 3.8 million tons of petroleum products (e.g., petrol and petrol-based) every year (MPEMR 2019).

In addition to non-renewable sources, there are renewable energy sources in Bangladesh. These include mainly solar, hydro and small wind. Off-grid solar home systems (SHSs) are very popular ways of providing basic access to electricity in the rural area of Bangladesh. By 2021 the government is planning to increase this capacity of about 220 MW (Khan 2019b). Recently implemented solar power generation projects include a 7.4 MWp grid connected solar PV power plant in Rangamati at the Kaptai hydropower station compound, a 28 MWp solar park at Cox's Bazaar, and 8 MW at Tetulia (BPDB 2019). According to the Sustainable and Renewable Energy Development Authority (SREDA) of Bangladesh, the total installed renewable generation capacity (off- and on-grid) is 309.91 MW[3], of which major installed capacities came from solar (75.98 MW) and hydro (230 MW). On the other hand, 2.90 MW, 0.63 MW and 0.40 MW capacities came from wind, biogas, and biomass, respectively. In addition, there are some agricultural farms, such as dairy and poultry, in which biogas plants are being set up that produce biogas for cooking and offer limited electricity generation. Furthermore, electricity generation from the biomass gasification method also has potential in Bangladesh. Similarly, there is a huge potential for energy generation from Municipal Solid Waste using suitable technologies (Khan and Kabir 2020).

3.3.1 NON-RENEWABLE SOURCES

3.3.1.1 Natural Gas

To date, 27 gas fields have been discovered in the country; of them, 20 gas fields are in production (Petrobangla 2019), two gas fields are not in production and seven are suspended from production. As of December 2018, the cumulative production was 15840.22 BCF from the 20 gas fields. The estimated remaining gas reserve was 10337.73 BCF (as of 31st December, 2018) for these 20 gas fields (Petrobangla 2019). In summary, by the end of 2018, 16.44 TCF gas had been produced from the 27 gas fields, and the estimated total gas reserve was 11.47 TCF (Petrobangla 2019).

---

[3] http://www.sreda.gov.bd/ (accessed on 29-Jan-2020).

A dramatic fourfold increase in gas consumption for the last 20 years was reported by Petrobangla. For instance, gas consumption in 1995–1996 was about 698 MMcfd and this increased to 2645 MMcfd in 2015–2016 (MPEMR 2018). The demand is even higher, as there is a large unmet demand and a need for the curtailment of supply. An analysis shows that it is likely that demand will continue to grow (Petrobangla 2019). Natural gas accounts for 68% of the total grid electricity generation, while the fertilizer factories are dependent on natural gas for feedstock. For example, about 12% (230 million cubic feet) of the nation's total daily gas production is used to produce fertilizers (MPEMR 2019). Natural gas has made a significant contribution to industrial growth in the country, as it provides fuel for heating and captive power generation at very favourable prices. While the whole nation has benefitted from this resource, about 7% of the population has directly benefitted from using piped natural gas for household purposes (MPEMR, 2019).

As it is almost the only indigenous source of commercial energy, the demand for natural gas has experienced very fast growth over the last three decades, often outstripping the supply. In 2016–2017, demand for gas in the country was about 3736 MMcfd, whereas supply was 2754 MMcfd, indicating a daily shortage of about 27% (MPEMR 2018). It is estimated that demand for natural gas will rise to about 4957 BCF by 2023. A total of 2960 MMscfd was produced as of December 2018 (Petrobangla 2019). Given the reduction in production and supply of gas against the increasing demand, the remaining reserved gas will be depleted soon. The sector-specific gas consumption in the 2017–2018 financial year is shown in Figure 3.2.

3.3.1.2 Coal

Coal could be an alternative primary source of fuel to natural gas for power generation. In Bangladesh, the reserve of coal (bituminous) is about 3300 million tons. It is estimated that the available coal might conveniently serve the country's

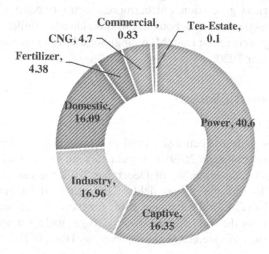

FIGURE 3.2 Sector-specific gas consumption (in %) in 2017–2018 financial year in Bangladesh. (Data source: Petrobangla 2019).

TABLE 3.1
Coal Production Scenarios of Last Ten Financial Years

Sl. No.	Financial Year	Production (MT)
1.	2017–2018	923,276.08
2.	2016–2017	1,160,657.81
3.	2015–2016	1,021,638.10
4.	2014–2015	675,775.50
5.	2013–2014	947,124.56
6.	2012–2013	854,803.85
7.	2011–2012	835,000.00
8.	2010–2011	666,635.39
9.	2009–2010	704,657.76
10.	2008–2009	827,845.06

Data source: BCMCL[4].

energy sector for 50 years. Its high level of heat generation capacity (11040 btu/lb) and low sulphur content (0.53%) has shown Bangladesh's coal to be of high quality (Petrobangla 2019). About 65% of the coal that is produced is used to generate electricity and the remaining 35% is used in brick fields and for other domestic purposes in the rural areas. In the 2017–2018 financial year, 923276 metric tons of coal were produced. At present, the Barapukuria coal mine is producing an average of 2500–3000 metric tons of coal per day and the measured and probable reserve is about 390 million tons (MPEMR 2018).The coal production from Barapukuria Coal Mining Company Limited (BCMCL) for the last ten financial years is depicted in Table 3.1. Barapukuria coal mine could produce one million tons of coal per year.

3.3.1.3 Petroleum

Petroleum is another energy source in Bangladesh containing kerosene, diesel and petrol. Petroleum products, such as diesel, petrol, octane furnace oil, etc. account for about 22% of the commercial energy supply in the country. Five million tons of petroleum products were imported by Bangladesh Petroleum Corporation in 2013–2014 financial year (MPEMR 2018). About 1.2 million metric tons of crude oil and 5.5 million metric tons of refined petroleum products are imported in Bangladesh per year.

About 49.40% of the total petroleum is consumed in the transportation sector. The agriculture sector consumes about 15.7% for irrigation. The remaining energy is used in power (26.94%), industry 4.86%, domestic (2.26%) sectors and other sectors 0.85%, as illustrated in Table 3.2. As rural electrification uses on-grid and off-grid solar PV, the dependency on petroleum (especially kerosene) is reduced by households. On the other hand, the use of petroleum in the transportation sector

---

[4] http://bcmcl.org.bd/production-statistics/ (accessed on 31-Jan-2020).

TABLE 3.2

Sector-Specific Petroleum Consumption in 2017–2018

Sector	Total (MT)	Percentage
Transportation	3432239	49.40%
Power	1871756	26.94%
Agriculture	1090903	15.70%
Industry	337903	4.86%
Domestic	156756	2.26%
Others	58779	0.85%
Total	6948336	100%

Data source: BPC[5].

and agricultural sector is increasing. If this trend continues, the demand for oil will increase to about 10 million tons by the year 2030.

3.3.2 RENEWABLE ENERGY

In 2041, total electricity demand for the high- and low-case scenarios will be about 82292 MW and 72000 MW, respectively (Power Division 2018). For the former case, the required renewable capacity will be 9400 MW, and for the latter case, it will be 7950 MW in 2041, considering there is a 10% renewable capacity in the generation fuel mix of the Bangladesh Power System Master Plan (PSMP). In the short-term plan, for example, up to 2021 for the high and low scenarios, the requirements will be 2800 MW and 2600 MW (Power Division 2018). To meet these renewable requirements for the generation fuel mix, the government of Bangladesh has already taken initiatives and year-wise planning has been initiated. A plan from 2018 to 2035 is depicted in Table 3.3. It can be seen that the planned 10% renewable capacity addition will be completed by 2035.

Although the planned generation of renewable power is dominated by solar, the main limitation of solar is the availability of the sun during the hours of peak demand at night. On the other hand, hydro is a proven electricity generation system that could easily meet peak demands, as shown in New Zealand, see (Khan, Jack, and Stephenson 2018) and (Khan 2019f). However, hydropotential in Bangladesh is very limited due to the geographical location of the country. Although this may be true, 5000 MW of electricity could be generated from hydro in Bhutan, Nepal and the North-Eastern part of India, which can be imported in Bangladesh through a cross-border power trading policy (Power Division 2018).

Studies show that biomass, such as waste to energy generation, has potential in Bangladesh, see (Swaraz et al. 2019) and (Khan 2019e). For example, wild date palm fruit pulp and sap have been found to have potential for bio ethanol production

---

[5] http://www.bpc.gov.bd/site/page/d6742c5d-2775-4014-b5f7-ad1cc5b1e6f5/ (accessed on 31-Jan-2020).

TABLE 3.3
Renewable Energy Generation Plan up to 2035

Year	Solar (MW)	Wind (MW)	Biomass (MW)	Total (MW)
2018	427	60	-	487
2019	1080	100	1	1181
2020	315	-	-	315
2021	100	150	-	250
2022	300	-	-	300
2027	-	50	-	50
2031	100	-	-	100
2035	-	150	-	150
Total	2322	510	1	2833

Data source: Power Division 2018.

(Swaraz et al. 2019). About 186 GWh/day electricity can be produced from a landfill gas recovery process in Bangladesh (Khan 2019e). Although waste (i.e., biomass) was found to have potential to generate energy, this source has received less attention in Bangladesh's PSMP.

3.4   ENERGY STATUS IN BANGLADESH

Electricity is an indispensable form of energy used for overall economic growth for any nation. One decade ago, an electricity crisis was one of the major concerns for Bangladesh's rural electrification and industrialization. Sufficient electricity supply was crucial for the country's gross domestic product (GDP) growth, as well as poverty reduction. The government of Bangladesh took the power sector as the thrust sector and called for investment in the sector. Donor agencies such as the World Bank and Asian Development Bank, and private companies responded by investing in the sector. Recently, electricity generation capacity has increased significantly in Bangladesh. For example, the installed generation capacity in 2013–2014 financial year was 9821 MW and this was increased to 18961 MW in 2018–2019, that is, the generation capacity has doubled within this short period of time, see (BPDB 2015) and (BPDB 2019). Urbanization is taking place at an unprecedented speed and producing a large demand for energy, see (Jones 1991) and (Sadorsky 2013). Bangladesh today is experiencing rapid transformation toward an urban society and the urbanization level has jumped dramatically from 8.78% to 27.66% between 1974 and 2011 (Islam 2018). In addition, the growth rate of the urban population on an average is 5.18, which is much higher than the growth rate of the rural population (1.36%), and the total population will live in urban areas by 2050, which indicates that continued urbanization in Bangladesh will promote the growth of national energy consumption, see (Islam 2018) and (Zhao and Zhang 2018). The activities of urban development require more energy because of the increasing amount of mobility, health, transportation and education services, see (Jones 1991), (Zhao and Zhang 2018), and (Salim, Rafiq, and Shafiei 2017). It is estimated that the population of Bangladesh will be around 230–250 million in 2050 with a zero-population growth rate. Household energy consumption will need more energy for freezing, cooking and other activities, see (Khan 2019c) and (Khan 2019d). Bangladesh is also experiencing industrialization and the trend of industrialization is increasing gradually, while, sequentially, energy consumption is intensifying at enormous speed.

3.5 ELECTRICITY SCENARIO IN BANGLADESH

In recent years, the electricity generation sector of Bangladesh has made significant progress. For instance, the annual capacity increment for the financial year of 2018–2019 was about 18.86% (BPDB 2019). The total installed capacity increased from 15,953 MW in 2017–2018 to 18,961 MW in the 2018–2019 financial year. However, the transmission loss increased from 2.76% to 3.15% for the same duration. This increase in generation was dominated by gas; about 57.37% electricity was produced using natural gas followed by furnace oil (25.16%) and diesel (7.23%). In contrast, only 1.37% of the generation was from renewable sources, including hydro and solar. The technology used for these types of electricity generation is depicted in Figure 3.3.

The total electricity generation from different renewable and non-renewable sources and their flow to the final distribution system is shown in the Sankey diagram in Figure 3.4. It can be seen from both Figures 3.3 and 3.4 that the dominating fuel in the electricity generation is gas followed by oil.

The number of consumers in different sectors, including domestic and commercial, is increasing remarkably. This is due to the population growth and industrialization in

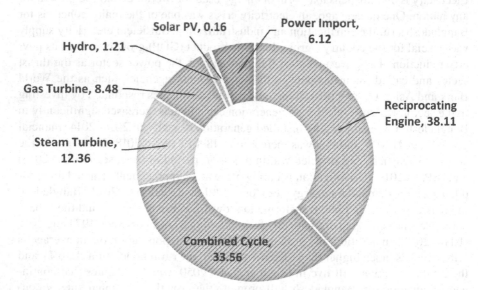

FIGURE 3.3   Electricity generation technologies (in %) used in the financial year 2018–2019 in Bangladesh. (Data source: BPDB 2019).

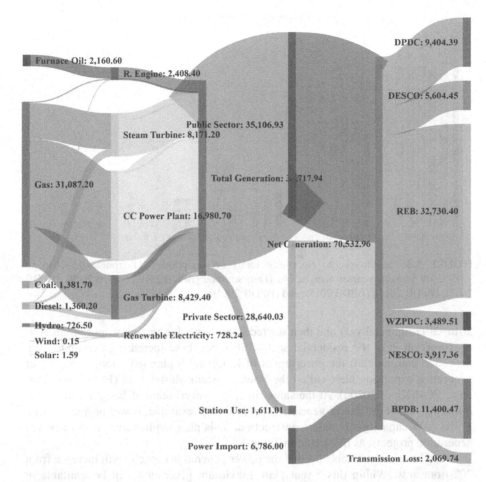

FIGURE 3.4 Sankey diagram illustrating energy flow from generation to distribution for the financial year 2018–2019. Note: Energy unit- GWh; R. Engine- Reciprocating Engine; CC Power Plant- Combined Cycle Power Plant; DPDC- Dhaka Power Distribution Company; DESCO- Dhaka Electric Supply Company; REB- Rural Electrification Board; WZPDC- West Zone Power Distribution Company; NESCO- Northern Electricity Supply Company; BPDB- Bangladesh Power Development Board. (Data source: BPDB 2019).

the developing countries, such as Bangladesh (Khan 2018). Per capita electricity generation and consumption increased from 382.18 kWh and 335.99 kWh in 2017–2018, to 425.92 kWh and 374.62 kWh in the 2018–2019 financial year, respectively (BPDB 2019). In the 2018–2019 financial year, the total installed capacity in the public and private sectors, and from imports were 9507, 8294 and 1160 MW, respectively.

Figure 3.5 clearly illustrates increased consumption in the domestic and commercial sectors in the 2018–2019 financial year, compared to the previous years. For the agriculture sector, a gradual decrease in consumption was observed. In contrast, the consumption in the 'others' sector has an increasing trend in nature. On the other hand, the consumption in the industrial sector reached its maximum in the

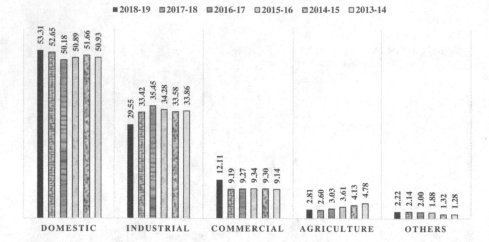

FIGURE 3.5 Sector-specific electricity retail consumptions (%) from 2013–2014 to 2018–2019 financial years in Bangladesh. (Data sources: (BPDB 2019), (BPDB 2018), (BPDB 2017), (BPDB 2016), (BPDB 2015), and (BPDB 2014)).

2016–2017 financial year and then started decreasing. Interestingly, more than 50% consumption is in the residential sector. To meet these increasing demands (e.g., domestic, commercial) the government of Bangladesh planned a long-term power generation expansion plan, called the Power System Master Plan (PSMP) for 2016 up to 2041(Khan 2020d). At the same time, the government of Bangladesh is also initiating short-term power generation projects. For example, power projects with a 17304 MW capacity are under construction. This plan implies five years of power generation projects, as illustrated in Table 3.4.

It can be seen from Table 3.4 that the power generation capacity will increase from 2020 onwards. Within this 5-year plan, maximum generation will be available in 2022, approximately a 68% increase from the previous year. In the following year, this increment will be 1.67%. All these yearly plans are dominated by the public sectors. In 2022, a big share will be added to the generation capacity through power imports from the neighboring countries. On the other hand, the long-term power generation plan

TABLE 3.4

Short-Term (5 Years) Power Generation Plan for Bangladesh

Sector (MW)	2019	2020	2021	2022	2023
Public	1463	1953	2217	870	2890
Private	2090	553	0	1942	1490
Import	0	0	340	1496	0
Total	3553	2506	2557	4308	4380

Data source: BPDB 2019.

includes 31000 MW and 57000 MW capacities against 27400 MW and 51000 MW demands in 2030 and 2041, respectively (BPDB 2019). In 2041, these planned generations will be available from 35% coal, 35% gas, 12% nuclear, 2% from renewable and others, and 16% from imports (Government of Bangladesh 2016), (BPDB 2019).

3.5.1 CHALLENGES AND FUTURE DIRECTIONS

The main challenge in the power generation expansion plan of Bangladesh is to implement the sustainable development goals imposed by the United Nations. These goals include: first, ensuring energy access for the total population of the country; second, introducing a considerable share of renewable energy into the generation fuel mix; third, importing power from neighboring countries; and, finally, related policy making for the energy sector.

To provide electricity access to the total population, the government has already prepared a long-term power generation expansion plan. Importantly, the plan is dominated by fossil fuels, such as coal and gas (Government of Bangladesh 2016). However, fossil fuels have limited resources; thus, if these resources are used continuously, they will be depleted after a certain time. Therefore, this is not a sustainable power generation plan (Khan 2019e).

To increase the renewable share in the electricity generation fuel-mix, the PSMP-2016 includes 10% generation from various sources such as solar, hydro and biomass. This generation includes an off-grid and on-grid electricity system. The generation potentials assessed by PSMP-2016 are depicted in Figure 3.6. Maximum generation could be achieved from solar, which is about 2680 MW. This includes solar parks, solar rooftops, solar home systems and solar irrigation systems technologies. The second highest potential was found to be the wind. The third potential source was estimated as waste to energy options, including energy generation from biomass, biogas, and municipal wastes. On the other hand, the least feasible options were small hydro and hybrid generation. Hybrid generation consists of mini- and micro-grid electricity systems, such as prosumerism (Khan 2019b). However, it is difficult and a real challenge to integrate these low capacities from off-grid systems into the main grid.

Power imports from neighboring countries could be a solution to this fossil fuel dominated and renewable constrained generation system. In addition, power imports also offer many other advantages, such as load management. For instance, the 'peak demand' period in one country might be an off-peak time in another neighboring country. Thus, the country that has available excess generation can transmit it to another country with a deficit in power (Khan 2020d). At present, Bangladesh is already importing power from India. In the 2018–2019 financial year, 6.12% of the total power was imported from India. PSMP-2016 also assessed that 3500–8500 MW hydropower potential is available and that this might be transmitted to Bangladesh from Nepal and North-West India (Government of Bangladesh 2016) Although this is true, there are a few concerns that must be taken into account. Firstly, as a rule of thumb, imported power should not exceed 10% of the total electricity requirement of the country, as dependence on more than this limit may pose energy security risk to the importing country. For example, if a big share of electricity is imported from a neighboring country and if an accident in the electricity system occurs in the source

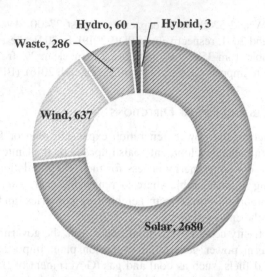

FIGURE 3.6 Renewable energy generation potential (in MW) in Bangladesh. (Data source: Government of Bangladesh 2016).

country, a complete black-out might occur in the destination country. On November 1, 2014, a nationwide blackout occurred in Bangladesh due to a problem that related to the power import from India[6]. Secondly, building high-voltage transmission lines between countries is costly and needs proper policy making. Sometimes, this might involve two or more countries. For example, electricity transmission from Nepal or Bhutan to Bangladesh must use India's transmission infrastructure. Therefore, proper policy making between countries must be guaranteed.

Fossil fuel depletion is the main challenge in the electricity generation sector of Bangladesh, as the electricity sector is dominated by fossil fuels. Table 3.5 shows the fossil fuel consumption in the power generation sector in the financial years 2008–2009 to 2018–2019. Of these consumptions, natural gas was the main fuel used for electricity generation. For instance, 40% of gas was consumed by the power generation sector in the 2017–2018 financial year.

Most importantly, the short-term power generation plan shows that gas demand is going to increase in the coming years. This forecast is depicted in Figure 3.7. Here it can be seen that the gas demand is going to increase by about 23% in 2021–2022 and 2022–2023, from the 2018–2019 financial year. Notably, gas demand for electricity generation is the highest in each year compared to other sectors.

The insufficiency of funding options for power generation projects is another challenge for least developed countries such as Bangladesh. In particular, this is the main concern for renewable electricity generation. Although the renewable energy generation technologies are cheaper nowadays, it is not in a range that can be afforded by all developing countries. In addition, all types of renewable generations are not suitable for every country.

Another challenge that exists almost in every developing country is that of power generation efficiency. In Bangladesh, for example, the average thermal efficiency of

---

[6] https://www.thedailystar.net/bangladesh-blackout-2014-48574 (accessed on 23-Jan-2020).

TABLE 3.5
Fuel Consumption in the Electricity Sector of Bangladesh

Year	Natural Gas (MMcft)	Furnace Oil (L)	Other Oil (L)	Coal (Ten Thousand Ton)
2008–09	1,61,007.68	90.26	112.81	0.47
2009–10	1,66,557.42	9.74	124.69	0.48
2010–11	150,031.41	118.78	137.66	0.41
2011–12	151,047.84	182.48	59.89	0.45
2012–13	175,944.51	266.11	34.97	0.59
2013–14	183,522.79	424.72	175.00	0.54
2014–15	180,765.64	378.13	291.06	0.52
2015–16	207,838.44	439.33	238.22	0.49
2016–17	215,894.52	512.56	347.98	0.59
2017–18	211,341.98	615.35	795.34	0.82
2018–19	273,920.59	480.06	372.50	0.57

Data source: BPDB 2019.

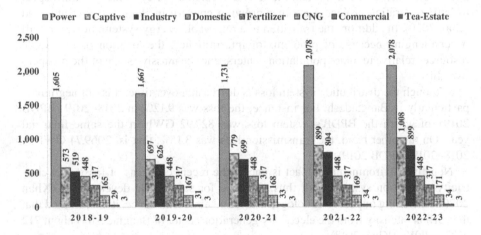

FIGURE 3.7 Short-term sector-specific future gas demand (in BCF) in Bangladesh. (Data source: Petrobangla 2019).

the power plants is about 38.4% (BPDB 2019). However, this efficiency can easily be improved up to 56% for gas, 42% for oil and 38% for coal-fired power plants. If these efficiencies can be implemented, generation will increase significantly. In addition, studies show that improving the efficiencies of the thermal power plants could reduce greenhouse gas emissions from 660–685 gCO_2-e/kWh to 460–520 gCO_2-e/kWh, see (Khan 2019a) and (Khan 2018).

Although there exists a huge potential for demand side management (DSM) in the developing world, it has received less attention in policy making and energy planning in Bangladesh. In designing proper DSM, there are four main challenges: cost of the DSM project; making sure appropriate technology is in place; related system complexity; and the consumers (i.e., the end users of electricity) (Khan 2019c). At the initial stage of DSM strategy selection for any country, 'energy-saving behaviour' could play a pivotal role among many different DSM schemes.

The cost of energy is a crucial factor for the electricity sector in Bangladesh. The government of Bangladesh is providing subsidies (see Figure 3.1) for the power sector and, as a result, this sector operates at loss. In the 2018–2019 financial year, for example, the operating revenue (including energy sales) was 34506 and expenses (e.g., fuel and maintenance cost) were 39553 Crore Taka. Thus, the loss was about 5047 Crore Taka (BPDB 2019). In the near future, if the government starts to reduce the subsidies, the consumer should pay all the expenses for the electricity. These costs will be collected by the distribution company through electricity tariffs (Khan 2019d).

A lack of appropriate investment policies is another key problem in Bangladesh. Most often, the generation technologies and resources that need to be prioritized to generate electricity over time receive less attention. Consequently, proper investment policies are not in place. For example, studies show that waste to energy (WtE) generation has potential in Bangladesh (Khan 2019e), (Khan and Kabir 2020). However, in PSMP-2016, WtE received less attention (Government of Bangladesh 2016). The lack of environmental assessment related policy making with respect to WtE generation project is another growing concern in Bangladesh (Kabir and Khan 2020). In addition, the transition to a renewable energy system, however, will be challenging because of problems of intermittency, the location of renewable resources relative to major population centers, and the massive scale of the prospective shift[7].

Although the distribution system loss is decreasing over time, it is not negligible, particularly in Bangladesh. For instance, the loss was 9.12% in 2018–2019 (BPDB 2019), of which the BPDB's system loss was 827.92 GWh in the same financial year. On the other hand, the transmission loss was 3.15%, that is, 2069.74 GWh in 2018–2019 (BPDB 2019).

Negative environmental impact is one of the recent concerns of fossil-fuel electricity generation systems and, thus, an issue for sustainable development (Khan 2020d) including impacts on ecology, water, and air. For instance, it is estimated that the carbon intensity from the electricity generation sector of Bangladesh is about 712 gCO_2-e/kWh (Khan 2018).

3.5.2 Present Scenario

The government of Bangladesh has significantly increased its installed power generation capacity, including power imports through cross-border trading and off-grid captive generation (Government of Bangladesh 2016). In view of the depleting proven reserve of its own natural gas, imported liquid fuel, liquefied natural gas (LNG) and imported power have been added to the fuel mix. Very soon imported coal and nuclear power will also be added to the fuel mix. Work on development of coal port and coal

---

[7] http://blogs.worldbank.org/developmenttalk/the-global-energy-challenge

trans-shipment terminal is progressing at Matarbari and Payra. Imported coal-based power plants are under construction at Matarbari, Maheshkhali, Banshkhali, Payra and Rampal, while Floating Storage and Regasification Units (FSRUs) and land-based LNG terminals are being set up at Maheshkhali, Matarbari, Payra and Kutubdia.

PSMP-2016 offers a long-term power generation plan, which includes the generation of 4,000 MW electricity from nuclear power; as such, the country is expected to have 2,400 MW of nuclear power by 2023. Initiatives are maturing for the import of more power from India, Nepal and Bhutan under bilateral and multilateral initiatives. Projects are in progress for expanding and modernizing the power and gas transmission and distribution system.

All these require a few thousand qualified and trained human resources (competent technical experts and managers). Bangladesh needs to adopt an affordable primary fuel mix without creating a situation of exclusive reliance on imported fuel. The country also needs to rationalize the fuel and energy utilization and the economic pricing of energy and power. It is hoped the government has done its homework, as hesitations and indecisions may trigger a crisis, as the global scenario relating to energy is also undergoing a major policy shift. The world is steadily moving away from fossil fuel. In the backdrop of the above, let us discuss energy and power sector challenges and priorities. In our opinion, challenges exist in six significant sectors, as shown in Figure 3.8.

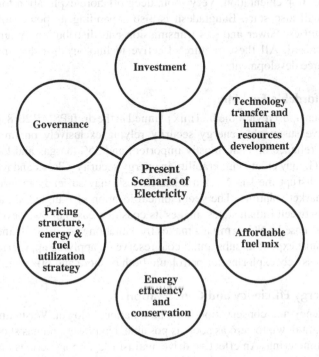

FIGURE 3.8 Present scenario for the electricity generation problem in Bangladesh.

3.5.2.1   Investment

In the paradigm shift of fuel choice with the world steadily drifting away from fossil fuel, Bangladesh is struggling to acquire the huge investment required for primary fuel exploration, the infrastructure development for importing primary fuel, the setting up of power plants, and the transmission and distribution infrastructure. Bangladesh has a negligible carbon footprint. As such, it has no obligation to limit emissions. Yet, being one of the most vulnerable countries to climate change impacts, it has to adopt expensive modern technology for energy generation from fossil fuels coal and LNG. During the last decade, the government successfully acquired loans from the Japanese, Chinese and Russian governments for major energy and power infrastructure development. A very attractive private sector power policy has helped develop the capability of local investors in setting up small- to medium-scale power generation plants. Apart from these, very little investments have been made in primary fuel exploration, mid- and downstream energy, and power infrastructure development.

3.5.2.2   Technology Transfer and Human Resources Development

Bangladesh has adopted several medium to large imported coal-based power generation projects. For these, the adoption of modern ultra-super critical technology and use of superior quality coal are essential. Bangladesh can use its quality coal. But it needs to be mined and transported from mines to power plants in an environment-friendly manner. Bangladesh is also implementing a nuclear power project using VVER-1200 third generation, plus Russian technology. Several LNG import projects are also under implementation. Very soon, deep offshore exploration for petroleum exploration will also start. Bangladesh is also expanding its power imports from regional countries. Power and gas transmission and distribution systems are also being modernized. All these require effective technology transfer and extensive human resource development.

3.5.2.3   Affordable Fuel Mix

The government must review the fuel mix planned in the draft PSMP 2018. Bangladesh cannot achieve sustainable energy security relying exclusively on imported fuel. Around 35% reliance on coal, mostly imported, and 35% on gas and LNG, mostly imported LNG, may create vulnerability in energy security. Global and regional geopolitics may disrupt the supply chain. Bangladesh may suffer from price shock in global fuel market volatility. The geographical location of Bangladesh, a river delta sandwiched between Indian states, makes its coastal areas shallow. Developing coal ports and land-based LNG terminals makes the situation very difficult and extensive. Bangladesh must exploit its substantial coal reserve by applying appropriate technology and aggressively exploring for petroleum both onshore and offshore.

3.5.2.4   Energy Efficiency and Conservation

Energy efficiency and conservation must be a national slogan. Waste and pilferage must be brought down to zero as soon as possible. No power and gas should be supplied without metering. An effective drive against illegal connections and enforced penalties for mischief mongers must be priorities in the operation of the power and energy sectors. SREDA and Bangladesh Energy Regulatory Commission (BERC) must be activated for auditing energy efficiency. All power and energy using appliances must be standardized and star rated. Fuel inefficient power plants, fertilizer factories and industries must be modernized with Balancing, Modernization, Rehabilitation, and Expansion (BMRE). Cogeneration and tri-generation must be the future options for fuel use.

3.5.2.5 Fuel Utilization Plan

Within a reasonable time, the government must announce fuel imports and a utilization plan. Investors must know what fuel will be supplied to which category of users at what price. LPG and piped gas price must be competitive. Similarly, the price of gas for compressed natural gas (CNG) and diesel price must also be competitive. We find no reason why domestic consumers and CNG cannot receive a gas supply if they agree on a price and every supply comes under pre-paid metering system. The BERC must be the sole authority for setting and regulating power and energy price until the Bangladesh market matures for deregulation.

3.5.2.6 Governance

Last but not least, effective governance is a major issue that the government must address with due attention. In relation to the energy and power sector, the government must restrict its role in law, acts, policy and regulation formulation, and in overseeing the operation of business. State-Owned Enterprises (SOEs) and companies formed under a company's act must run business without undue interference by the government. The BERC can police the enforcement of and compliance to acts, policies and regulations through auditing and stakeholders' engagement. The energy and power sectors are technically intensive areas. Here, non-technical officials in senior positions frustrate development. The non-transparent Speedy Supply of Power and Energy (Special Provision) Act must be rescinded within months of the government's assuming office. All new contracts and agreements in the power and energy sectors must follow the national procurement policy. The energy and power sectors are strategic areas. There must not be any political bias or victimization of officials for political faiths and beliefs. Bangladesh has produced many quality energy and power sector professionals over the years. Most of them have left the country and are contributing abroad. The sector suffers from a lack of competence and efficiency. Many of them were politically victimized and knocked out by an evil nexus of corrupt syndicates. Unless the government identifies, assesses and takes speedy action in curing the ills of energy and power sectors, the energy security essential for seamless economic development may remain a distant dream.

3.5.3 Energy and Environment

Electricity is one of the predominant components in enhancing the economic activities of a country. Bangladesh is not an exception to this trend. This has been reflected in the long-term power generation plan of the country. In Vision 2041, the government of Bangladesh has indicated the desire to be a developed country. To achieve this vision, economic development is indispensable and for this, the development of a power-generation capacity of the country is a prerequisite (Saha and Lopa 2019). Importantly, the government has already initiated a power generation expansion plan for this scenario. This plan includes approximately 35% gas and 34% coal-fired electricity generation by 2041 (Khan 2019e).

Bangladesh is one of the worst countries in the world in regard to environmental pollution. Fossil-fuel electricity generation will increase this pollution (e.g., air pollution) in the near future. The burning of fossil fuels for electricity generation will emit toxic elements, such as SO_x, NO_x, CO and small particulate matter (PM) including PM_{10} and $PM_{2.5}$. For example, in a recent study, it was reported that a gas-fired power plant in Bangladesh with an efficiency of 21.76% in carbon intensity was estimated as being 930.84 gCO_2-e/kWh (Khan 2018). Even the emission was found to be almost double (1752.02 gCO_2-e/kWh) for an oil-fired power plant with an efficiency of 15.97% (Khan 2018).

In addition, the transportation of fossil fuels to the plant will also add a similar type of pollutants to the air/environment. If imported coal is carried to the power plant by marine vessels, this will have additional negative environmental impacts to the water route and its surrounding. These pollutants create short- and long-term health issues such as breathing problems and heart disease. Moreover, the increase of SO_2 in the environment will result in acid rain, which might damage crops, forests and soil; overall, it will degrade the ecosystem.

Although each and every electricity generation project must fulfil social and environmental requirements, it is a matter of great concern that, most often, these impact assessments (i.e., Environmental and Social Impact Assessment (ESIA)) are not conducted in accordance with the standard guidelines in the developing world (Khan 2020a). For any new power generation project, either fossil-fuelled or renewable, proper impact assessments should be conducted (Kabir and Khan 2020).

3.6 POLICY RECOMMENDATIONS

Bangladesh has the vision to become a developed nation by 2041; thus, it is indispensable to follow the rules and regulations that are imposed by international organizations such as the United Nations (UN). One such criterion is the sustainable development goals (SDGs) of the UN. Providing access to electricity for any country's total population is one of the targets of SDG goal 7. Therefore, the government of Bangladesh has initiated the PSMP to achieve the goal. Vision 2041 goals for the power sector are summarized in Table 3.6. However, in addition to this, the government must ensure that certain portions of electricity generation from renewable sources be used to reduce the use of fossil fuels that generate electricity.

There are some recommendations that might be helpful for Bangladesh to achieve its Vision 2041 for the power sector. These are:

*   Increasing demands from different sectors must be followed in devising an effective generation expansion plan.
*   Generation capacity should be built in such a way that the reserve capacity be 15–20% higher than the actual demand required to meet immediate peak demands.

TABLE 3.6
Vision 2041 for the Power Sector of Bangladesh

Parameters	Year-2018	Year-2041
Power shortage	Seasonal	None
Per capita generation (kWh)	464	2100
Average plant's thermal efficiency (%)	30	50
Frequency fluctuation (Hz)	±1.5	±0.2
Power demand (MW)	10500	72000
Generation capacity (MW)	13500	79500
Generation from gas/LNG (MW)	8700	34000
Generation from coal (MW)	436	25500
Generation from liquid fuel (MW)	4500	1800
Generation from nuclear (MW)	0	5500
Import (MW)	1160	12000
Renewable generation (MW)	300	7900

Data source: Power Division 2018.

- There must be coordination between the generation, transmission and distribution expansion plan.
- The power transmission network must be connected in such a way that maximum redundancy can be guaranteed. In addition, the transmission and distribution network should be designed in such a way that they can be separated into different zones so that national blackouts can be avoided.
- Demand side management, energy efficiency and conservation measures must be ensured to manage the peak demands and the proper utilization of the network infrastructure.
- When importing power from the neighboring countries, it must be ensured that the imported power should not exceed 15% of the total national demand. In addition, no more than 10% power should enter through a single point, to avoid grid instability. Therefore, a cross-border electricity trading policy must be developed. This power trading policy must not only consider power imports from other countries, but also the exportation of power to those countries at the time of surplus generation.
- The electricity generation infrastructure developed by the government must not be below 60% in order that the power market can be well controlled by the government when considering a country's needs.
- The percentage of renewable generation needs to be increased according to all the possible sources.
- The long-term generation expansion master plan must be revised at least every five years. The sustainability assessment (using the standard method (Khan 2020b)) of this plan should also be conducted in relation to a sustainable future electricity generation sector.

3.7   CONCLUSIONS

In this chapter, an overview of the power generation sector of Bangladesh is discussed. The main topics included the primary energy sources that are used to generate electricity and their current status and future prospects. Bangladesh has the vision to become a developed nation by 2041; thus, the government of Bangladesh has taken initiatives to fulfil the requirements relating to this. The Power System Master Plan is one of these steps; as a part of this plan, 79500 MW generation capacity will be built by 2041. However, there are many challenges that need to be faced by the government to achieve this plan, such as an improvement from 30% to 50% in thermal power plants' average efficiency. Another problem is the depletion of fossil fuels and lack of suitable renewable generation sources. For instance, solar has huge potential in Bangladesh, but is unable to meet the peak demand due to its lack of availability. The same might be true for the wind due to the uncertainty of generation. On the other hand, hydro could be a solution to this, but Bangladesh has very limited options for further hydrodevelopment. However, cross-border electricity trading might be a potential option that could solve the fossil fuel dependence. At the same time, electricity imports from Bhutan and Nepal should be encouraged as the source of generation is hydro in these two countries. This will help to establish a regional sustainable development, which in turn will underpin the processes undertaken to achieve the United Nations' sustainable development goals.

REFERENCES

BPDB. 2014. "Annual Report 2013-2014." Bangladesh Power Development Board. Dhaka, Bangladesh: Bangladesh Power Development Board. https://www.bpdb.gov.bd/bpdb_new/index.php/site/new_annual_reports.

BPDB. 2015. "Annual Report 2014-2015." Bangladesh Power Development Board. Dhaka, Bangladesh. http://www.bpdb.gov.bd/.

BPDB. 2016. "Annual Report 2015-2016." Bangladesh Power Development Board. Dhaka. http://www.bpdb.gov.bd/bpdb/.

BPDB. 2017. "Annual Report 2016-2017." Bangladesh Power Development Board. Dhaka. http://www.bpdb.gov.bd/bpdb/.

BPDB. 2018. "Annual Report 2017-2018." Bangladesh Power Development Board. Dhaka. http://bpdb.gov.bd/bpdb_new/index.php/site/new_annual_reports.

BPDB. 2019. "Annual Report 2018-2019." Bangladesh Power Development Board. Dhaka, Bangladesh. http://www.bpdb.gov.bd/bpdb_new/index.php/site/new_annual_reports.

EECMP. 2015. "Energy Efficiency and Conservation Master Plan up to 2030." Sustainable and Renewable Energy Development Authority. Dhaka. http://sreda.gov.bd/files/EEC_Master_Plan_SREDA.pdf.

Government of Bangladesh. 2016. "Power System Master Plan 2016." Dhaka, Bangladesh. https://powerdivision.gov.bd/site/page/f68eb32d-cc0b-483e-b047-13eb81da6820/Power-System-Master-Plan-2016.

Halder, P. K., N. Paul, M.U.H. Joardder, and M. Sarker. 2015. "Energy Scarcity and Potential of Renewable Energy in Bangladesh." *Renewable and Sustainable Energy Reviews* 51: 1636–49. https://doi.org/10.1016/j.rser.2015.07.069

Islam, Nazrul. 2018. "Urbanisation in Bangladesh: Recent Trends and Challenges." *The Daily Sun*, October 24, 2018. https://www.daily-sun.com/post/345303/2018/10/24/Urbanisation-in-Bangladesh:-Recent-Trends-and-Challenges.

Islam, Saiful, and Md. Ziaur Rahman Khan. 2017. "A Review of Energy Sector of Bangladesh." *Energy Procedia* 110: 611–18. https://doi.org/10.1016/j.egypro.2017.03.193.

Jones, Donald W. 1991. "How Urbanization Affects Energy-Use in Developing Countries." *Energy Policy* 19 (7): 621–30. https://doi.org/10.1016/0301-4215(91)90094-5.

Kabir, Zobaidul, and Imran Khan. 2020. "Environmental Impact Assessment of Waste to Energy Projects in Developing Countries: General Guidelines in the Context of Bangladesh." *Sustainable Energy Technologies and Assessments* 37 (100619): 1–13. https://doi.org/https://doi.org/10.1016/j.seta.2019.100619.

Khan, Imran. 2018. "Importance of GHG Emissions Assessment in the Electricity Grid Expansion toward a Low-Carbon Future: A Time-Varying Carbon Intensity Approach." *Journal of Cleaner Production* 196: 1587–99. https://doi.org/10.1016/j.jclepro.2018.06.162.

Khan, Imran. 2019a. "A Temporal Approach to Characterizing Electrical Peak Demand: Assessment of GHG Emissions at the Supply Side and Identification of Dominant Household Factors at the Demand Side." University of Otago. http://hdl.handle.net/10523/9185.

Khan, Imran. 2019b. "Drivers, Enablers, and Barriers to Prosumerism in Bangladesh: A Sustainable Solution to Energy Poverty?" *Energy Research & Social Science* 55: 82–92. https://doi.org/10.1016/j.erss.2019.04.019.

Khan, Imran. 2019c. "Energy-Saving Behaviour as a Demand-Side Management Strategy in the Developing World: The Case of Bangladesh." *International Journal of Energy and Environmental Engineering* 10 (4): 493–510. https://doi.org/10.1007/s40095-019-0302-3.

Khan, Imran. 2019d. "Household Factors and Electrical Peak Demand : A Review for Further Assessment." *Advances in Building Energy Research*, 1–33. https://doi.org/10.1080/17512549.2019.1575770.

Khan, Imran. 2019e. "Power Generation Expansion Plan and Sustainability in a Developing Country: A Multi-Criteria Decision Analysis." *Journal of Cleaner Production* 220: 707–20. https://doi.org/10.1016/J.JCLEPRO.2019.02.161.

Khan, Imran. 2019f. "Temporal Carbon Intensity Analysis: Renewable versus Fossil Fuel Dominated Electricity Systems." *Energy Sources, Part A: Recovery, Utilization and Environmental Effects* 41 (3): 309–23. https://doi.org/10.1080/15567036.2018.1516013.

Khan, Imran. 2020a. "Critiquing Social Impact Assessments: Ornamentation or Reality in the Bangladeshi Electricity Infrastructure Sector?" *Energy Research & Social Science* 60 (101339): 1–8. https://doi.org/10.1016/j.erss.2019.101339.

Khan, Imran. 2020b. "Data and Method for Assessing the Sustainability of Electricity Generation Sectors in the South Asia Growth Quadrangle." *Data in Brief* 28 (104808): 1–8. https://doi.org/10.1016/j.dib.2019.104808.

Khan, Imran. 2020c. "Impacts of Energy Decentralization Viewed through the Lens of the Energy Cultures Framework: Solar Home Systems in the Developing Economies." *Renewable and Sustainable Energy Reviews* 119 (109576): 1–11. https://doi.org/10.1016/j.rser.2019.109576.

Khan, Imran. 2020d. "Sustainability Challenges for the South Asia Growth Quadrangle: A Regional Electricity Generation Sustainability Assessment." *Journal of Cleaner Production* 243 (118639): 1–13. https://doi.org/10.1016/j.jclepro.2019.118639.

Khan, Imran, Michael W. Jack, and Janet Stephenson. 2018. "Analysis of Greenhouse Gas Emissions in Electricity Systems Using Time-Varying Carbon Intensity." *Journal of Cleaner Production* 184: 1091–1101. https://doi.org/10.1016/j.jclepro.2018.02.309.

Khan, Imran, and Zobaidul Kabir. 2020. "Waste-to-Energy Generation Technologies and the Developing Economies: A Multi-Criteria Analysis for Sustainability Assessment." *Renewable Energy* 150: 320–33. https://doi.org/https://doi.org/10.1016/j.renene.2019.12.132.

MPEMR. 2018. "Gas Sector Master Plan Bangladesh 2017." Dhaka, Bangladesh. https://mpemr.gov.bd/assets/media/pdffiles/Bangladesh_GSMP_Final_Report.pdf.

MPEMR. 2019. "Energy and Mineral Resources Division." Ministry of Power, Energy and Mineral Resources. 2019. https://mpemr.gov.bd/power/details/80.

Petrobangla. 2019. "Annual Report 2018." Bangladesh Oil, Gas and Mineral Corporation. Dhaka, Bangladesh. https://petrobangla.org.bd/?params=en/annualreport.

Power Division. 2018. "Revisiting PSMP 2016." Dhaka, Bangladesh. https://powerdivision.portal.gov.bd/sites/default/files/files/powerdivision.portal.gov.bd/page/4f81bf4d_1180_4c53_b27c_8fa0eb11e2c1/Revisiting PSMP2016%28full report%29_signed.pdf.

Sadorsky, Perry. 2013. "Do Urbanization and Industrialization Affect Energy Intensity in Developing Countries?" *Energy Economics* 37: 52–59. https://doi.org/10.1016/j.eneco.2013.01.009.

Saha, Polin Kumar, and Irin Afrin Lopa. 2019. "Environmental Demand of Energy Sector in Bangladesh." *The Independent*, August 23, 2019. http://www.theindependentbd.com/printversion/details/212371.

Salim, Ruhul, Shuddhasattwa Rafiq, and Sahar Shafiei. 2017. "Urbanization, Energy Consumption, and Pollutant Emission in Asian Developing Economies: An Empirical Analysis." *ADBI Working Paper Series*. 718. April. https://www.adb.org/publications/urbanization-energy-consumption-pollutant-asian-developing-economies.

Swaraz, A.M., Mohammed A. Satter, Md. Mahfuzur Rahman, Mohammad Asadullah Asad, Imran Khan, and Md. Ziaul Amin. 2019. "Bioethanol Production Potential in Bangladesh from Wild Date Palm (*Phoenix Sylvestris Roxb.*): An Experimental Proof." *Industrial Crops and Products* 139 (111507): 1–9. https://doi.org/10.1016/j.indcrop.2019.111507.

World Bank. 2018. "Rural Population (% of Total Population) - Bangladesh." https://data.worldbank.org/indicator/SP.RUR.TOTL.ZS?locations=BD.

Zhao, Pengjun, and Mengzhu Zhang. 2018. "The Impact of Urbanisation on Energy Consumption: A 30-Year Review in China." *Urban Climate* 24: 940–53. https://doi.org/10.1016/j.uclim.2017.11.005.

4 100% Renewable Energy Strategy: Bhutan's Energy and Environmental Perspective

Hari Kumar Suberi

CONTENTS

4.1 Introduction 63
    4.1.1 Background of Bhutan..64
        4.1.1.1 Energy Demand and Supply Historical Development.........65
        4.1.1.2 Current Scenario, Demand and Supply 66
    4.1.2 Scenario 1: Enabling Policy..71
        4.1.2.1 Pricing Mechanism 71
        4.1.2.2 Energy Trading 72
    4.1.3 Scenario 2: Optimization and Demand Projection............................73
        4.1.3.1 Industrial Growth..75
        4.1.3.2 Livelihood Improvement...75
        4.1.3.3 Efficiency Improvement..76
    4.1.4 Scenario 3: Alternative Energy Potentials.......................................78
        4.1.4.1 Hydropower Potential 80
        4.1.4.2 Solar, Wind and Small Hydro...81
        4.1.4.3 Biomass Potentials 83
        4.1.4.4 Levelized Cost of Renewable Energy 85
        4.1.4.5 Fossil Deposit in Bhutan 86
4.2 Environmental Impact 87
4.3 Conclusion 91
References...91

4.1 INTRODUCTION

Bhutan's ambitious vision to remain carbon-negative is uniquely challenged unlike many other developing countries in the SAARC region. De-carbonization strategy is based on forest-conservation and hydropower as the main source of electrical energy supply, which is inevitably linked to the adverse impacts of climate change. As of now, only 5% of the total hydro potential is utilized for electricity generation, whereas 75% of the generation is exported to India. By the end of 2025 hydro generation capacity is expected to be more than 5,000 MW harnessing about 22% of total available potential. Despite the enormous potential, as Bhutan's transport fuel requirement is 100% import-dependent, the national energy balance currently remains questionable owing to the fluctuating fuel price and limited access to direct fuel supply.

While most countries in the SAARC region experience the energy supply deficit, Bhutan faces the challenge of how to optimize the energy demand, improve the efficiency and diversify the energy mix through renewable sources. Optimization strategy of energy access has ledthe government to remove the already established solar home system to all grid connected hydro-based electricity supply. The future scenario of this accessibility right is challenged by rural-urban migration leaving 100% rural electrification infrastructure operation and maintenance cost to remain increasingly expensive. Lack of data availability and technical capacity has also caused the energy diversification limited to single electrical energy source although thermal sources such as solar and biomass are viable in Bhutan. The qualitative analysis carried out on the available renewable energy (RE) sources indicates that Bhutan has about 9 GW of solar installed capacity with capacity factor of up to 30% for 4–5 KWh/m^2 solar radiation, 23.4 GW of hydro installed capacity from its 127.3 km^3 of ice reserve and with the average firm power capacity factor of 50%, where wind and biomass potentials are yet to be explored. The wind speed is recorded to be as high as 6 m/s. Biomass from cattle waste is seen viable for rural household cooking energy demand. However, more detailed data are required to estimate the biomass potentials. Combined potential of these RE sources have the capacity to mitigate the adverse impact of climate change. Thus the ambitious vision to remain a carbon-negative economy can be supported by these alternative energy sources and the supporting policies to implement them.

4.1.1 BACKGROUND OF BHUTAN

Landlocked and sandwiched between India and China with the land area of 38,394 km^2 and located in eastern Himalayas at latitude 27.5142°N and longitude 90.4336°E, Bhutan is a small independent nation state with approximately 735,553 people (NSB 2018). The country's development model is based on the well-being measure of its citizen defined in its unique development philosophy of gross national happiness (GNH). The holistic development approach with the overarching four pillars: sustainable and equitable development, environmental conservation, preservation and promotion of cultural values and good governance have already localized the global sustainable development goals (Harris 2017). Based on these national baseline conditions and philosophical development approach, all citizens are also ensured with 100% access to clean electrical energy and with 70% of its land under forest cover, it is currently envisioned to remain carbon neutral which is in sync with the GNH philosophy and climate action commitment (Yangka 2018).

While there is no clear data record on historical development of energy sources in Bhutan, due to its landlocked geography and forest coverage, it is worthwhile to assume the initial energy source as biomass especially the fuel wood and limited fossil fuel sources such as kerosene. However, we need more detailed data for defining a historical baseline scenario. The first exposure to the modern form of energy supply was electricity generation by diesel generator with the installed capacity of 256 KW in 1966 in Phuntsholing, the border town of Bhutan bordering with India (Tshering and Tamang 2004). In those early days, energy sector development was fully undertaken by the Ministry of Trade under the Department of Energy (DoE) which is now bifurcated into the Department of Renewable Energy (DRE), Department of Hydropower and Power Systems, Department of Hydro Met Services, Bhutan Electricity Authority(BEA), Druk Green Power Corporation (DGPC), and Bhutan Power Corporation (BPC). These entities aretoday generating, distributing and managing more than 1,615 MW electrical power installed capacity and planning the future energy potential (DRE 2015). The import of fossil fuel is mandated by theDepartment of Trade. More detailed energy data and institutional arrangement can be found in Bhutan energy data directory updated from 2005 and in 2015 by the DRE (DRE 2015).

4.1.1.1   Energy Demand and Supply Historical Development

Energy demand and supply has very strong influence on energy security both in terms of energy import and export. The security of energy supply is a multidimensional concept for both policy choice and planning purpose that will depend on availability, affordability and accessibility (Kocaslan 2014). It can also be assumed that the energy crisis since 1970 has not affected the energy supply scenario in Bhutan (Barsky and Kilian 2004). This could be partially due to late exposure to a modern form of energy which dates back to 1966 with only one 256 KW diesel generator (Tshering and Tamang 2004). Bhutan is, however, dependent on India for fossil fuel import and all regulation on fossil fuel in India will strongly affect the fuel price in Bhutan.

There is no scientifically documented information on how and when energy supply in Bhutan started. Probably the statistics on the use of forest biomass indicates that the first formal way of energy supply was fuel wood by the Department of Forests. With the establishment of the Department of Forests in 1952, the use of fuel wood is regulated and it is mandated that 60% of Bhutanese land be under forest cover, and as of 2018 the forest cover is 71% (MoAF 2018). Today this forest resource is empowered to the community as a community forest and, to date, there are 777 community forests under the land area 90,318.77 ha (MoAF 2018). Dominance of fossil energy sources at the household level was relatively negligible. The only fossil energy sources used are kerosene for lighting purposes in very few communities. Pine resin as a local fuel was also common in the absence of kerosene. In 2013 even when the rural electrification project was implemented, the national energy share of 70% is from forest biomass (Wangchuk, Siebert and Belsky 2013). For the transport sector energy supply, the rough baseline can be found from 1960 where first road construction work started (MoWHS 2015). However, there is no clear data record on how much fossil energy was supplied and how it was procured.

The energy supply and security has its root in the historical development philosophy and baseline condition in Bhutan. Both sustainable energy sources and fossil fuel import dependency are relatively recent phenomena. The commissioning of the first bilateral energy project in Chhuka with the installed capacity 336 MW hydro is the initial phase of national energy supply and international energy export to India (DGPC 2019a). The energy cooperation and mutual agreement have given a unique diplomatic tie between Bhutan and India.

4.1.1.2 Current Scenario, Demand and Supply

The current scenario constitutes the electrical energy originating from hydropower which is used for both thermal and electrical load and very recently also for fuel substitute in the transport sector. Including industries, institutions, households in rural and urban, and outdoor lighting, as of 2018, there are 192,859 consumers supplied with domestically generated electricity from a hydro source (BPC 2018). The transport sector energy demand is increasingly faced with challenges in terms of energy security and environment pollution owing to the growing motorization trend. As of 2019 there are 104,963 vehicles compared to 100,000 vehicles in 2018 (RSTA 2019). Even after 100% rural electrification work, few rural communities still depend on fuel wood for thermal energy, especially for heating in winter and for cooking purpose. Other fossil fuel imports include the liquefied petroleum gas (LPG) for cooking, diesel and petrol mostly for transport energy demand, and aviation fuel for international transport (NSB 2018).

Table 4.1 shows the fossil fuel import in Bhutan from India since 2013. The trade statistics shows high speed diesel (HSD), petrol, kerosene, liquefied petroleum gas (LPG), lubricating oil and petroleum dominate the fossil fuel imports. The kerosene subsidy/industrial (S/I) and kerosene jet fuel (J) represent the demand for

TABLE 4.1
Fuel Imports, in Kilo Liters (KL)/Metric Ton (MT)

Fuel Type	GJ/MT	(t C/TJ)	2013	2014	2015	2016	2017
Diesel	43.33	20.20	126,120.29	117,615.80	126,139.40	133,851.27	144,620.70
Petrol	44.80	18.90	30,272.20	31,458.20	33,846.60	36,214.80	39,119.50
Kerosene	44.75	19.60	4,966.01	24.00	48.00	24.01	12.00
Kerosene(S)	44.75	19.60	48.00	5,685.00	4,599.00	4,815.00	4,328.00
kerosene (I)	44.75	19.60	0.00	21.00	12.00	24.00	0.00
Kerosene(J)	44.59	19.50	3,168.00	3,585.00	3,291.00	3,294.00	3,982.00
LPG	47.31	17.20	6,719.48	7,056.18	7,289.73	7,593.30	8,078.90
Lubricating oil	40.19	20.00	2,573.47	2,512.93	3,378.26	2,478.56	2,239.17
Furnace oil	00.00	00.00	721.58	473.35	1,302.20	1,425.29	1,688.01
Light oil diesel	43.33	20.20	204.00	3,698.00	136.00	36.00	00.00
Petroleum bitumen	40.19	22.00	5,594.84	7,563.67	11,778.43	11,707.12	15,663.27
Other waste oil	40.19	20.00	218.28	172.03	5.40	10.79	140.38
Other containing biodiesel	40.19	20.00	58.29	2.51	00.00	00.00	00.00

Source: Compiled from NSB and IPCC.

TABLE 4.2
Electricity Generation and Supply, in GWh

Hydro Energy	2013	2014	2015	2016	2017	2018
Generation	7,531.45	7,147.10	7,381.60	7,573.84	7,248.92	6573.98
Domestic supply	1,901.73	2,064.30	2,142.17	2,084.69	2,317.28	2,454.31
Export	5,648.23	5,179.26	5,307.64	5,484.00	5,068.48	4,053.59
Import	-108.19	-187.37	-158.46	-110.64	-207.89	-300.34

Source: DGPC 2018.

commercial and aviation purposes. The energy content of these fossil fuel and their emission are adopted from the Intergovernmental Panel for Climate Change (IPCC) default values (IPCC 1996) and the United Nations (UN) manual (UN manual, n.d.). The standard units for energy are Giga Joule/Metric Ton (GJ/MT) and ton Carbon/ Terajoule (tC/TJ) as shown in Table 4.1. The imports have been increasing especially for diesel, petrol, bitumen and kerosene jet fuel.

Table 4.2 shows the national electrical energy balance from 2013 to 2018. The generation has been decreasing in general, the domestic demand has increased, export volume has also decreased, whereas the import has increased over the years. With the current installed capacity of 1,615 MW from 5 major power plants (DGPC 2019b), the power plants are on average running at capacity factor of 51.2%. The small-scale hydro is excluded from this list, as it will be discussed in alternative energy scenario 3.

Figure 4.1 shows the electrical energy load profile with an average monthly peak load for 6 months starting from January 2017 until June 2019. The x-axis shows the time of the day, and the y-axis shows monthly variation of peak load in Mega Watt (MW). The maximum peak load was recorded on 27th Dec., 2018 at 18:00 hr peak power demand of 399.35 MW (BPC 2019). The load profile and peak demand provide a better decision for the lean season energy-system planning. The peak load is also seen to be higher from 7 hr to 12 hr and highest from 18 hr to 23 hr.

Table 4.3 and Figure 4.2 show the power generation for 3 years, the monthly generation variation compiled from the BPC annual reports (BPC 2018) (BPC 2017) (BPC 2016). From Table 4.3 and the load profile Table 4.2, it shows why there is a need for import of electrical energy in the lean season especially in the winter months where peak load is increasing and generation is decreasing. This is critical in climate-energy era of the 21st century, where the lean seasons will increasingly face the challenge, owing to a longer dry season and low precipitation levels, the very common consequences of climate change.

Table 4.4 and Table 4.5 show the energetic value of the fossil fuel imported and the possible carbon dioxide (CO_{2e}) emission based on IPCC default values. The oxidation factor of 99% is used for all fuel types based on the IPCC default value (IPCC 1996). The energetic value of fuel imported has increased from 7,852.762 TJ in 2013

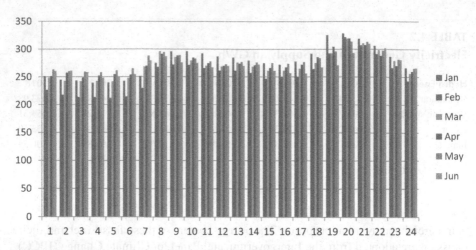

FIGURE 4.1 Electrical energy load profile (Source: Compiled from BPC data (BPC 2019), author's computation).

TABLE 4.3
Monthly Peak Power Capacity, in MW for 3 Years
(Source: BPC Data, BPC 2019)

Year		Jan	Feb	Mar	Apr	May	Jun	Jul	Aug	Sep	Oct	Nov	Dec
2018	Max	498.1	487.4	534.2	649.5	1228.3	1752.0	1718.5	1727.0	1724.0	1326.3	714.7	643.0
	Min	214.1	168.2	107.8	185.7	315.3	461.6	1089.6	1239.6	392.6	453.5	281.3	230.4
2017	Max	572.7	516.4	708.9	1214.5	1688.3	1751.8	1720.5	1725.3	1724.3	1665.7	847.5	568.4
	Min	203.6	211.4	231.5	186.9	321.2	345.4	1163.3	1100.7	850.7	740.9	445.1	296.6
2016	Max	398.2	514.6	605.8	871.0	1280.3	1740.3	1730.3	1723.9	1724.0	1724.4	1153.3	805.3
	Min	236.4	171.7	186.7	135.5	310.5	454.9	705.7	928.8	1067.3	1009.9	512.7	261.4

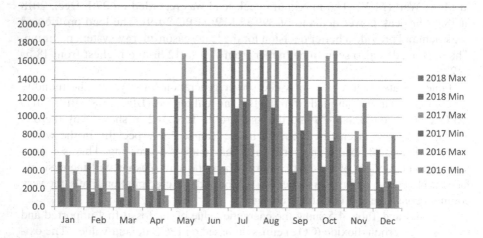

FIGURE 4.2 Monthly peak power capacity in MW for 3 years (Source: Compiled from BPC ata, author's compilation).

TABLE 4.4
The Energetic Value of Fossil Fuel, in Terajoule (TJ)

Fuel Type	2013	2014	2015	2016	2017
Diesel	5,464.79	5,096.29	5,465.62	5,799.78	6,266.41
Petrol	1,356.19	1,409.33	1,516.33	1,622.42	1,752.55
Kerosene	222.23	1.07	2.15	1.07	0.54
Kerosene(S)	2.15	254.40	205.81	215.47	193.68
kerosene (I)	0.00	0.94	0.54	1.07	0.00
Kerosene(J)	141.26	159.86	146.75	146.88	177.56
LPG	317.90	333.83	344.88	359.24	382.21
Lubricating oil	103.43	100.99	135.77	99.61	89.99
Furnace oil	0.00	0.00	0.00	0.00	0.00
Light oil diesel	8.84	160.23	5.89	1.56	0.00
Petroleum bitumen	224.86	303.98	473.38	470.51	629.51
other waste oil	8.77	6.91	0.22	0.43	5.64
Other containing biodiesel	2.34	0.10	0.00	0.00	0.00
Total	7,852.76	7,827.95	8,297.32	8,718.05	9,498.09

Source: Computation Based on IPCC Default Values.

TABLE 4.5
Carbon Dioxide Emission, in Metric Ton (MT)

Fuel Type	2013	2014	2015	2016	2017
Diesel	404,354.18	377,087.94	404,415.45	429,140.47	463,668.33
Petrol	93,890.30	97,568.72	104,976.43	112,321.48	121,330.51
Kerosene	15,954.88	77.11	154.22	77.14	38.55
Kerosene(S)	154.22	18,264.87	14,775.75	15,469.72	13,905.07
kerosene (I)	0.00	67.47	38.55	77.11	0.00
Kerosene(J)	10,090.07	11,418.21	10,481.82	10,491.38	12,682.66
LPG	20,028.76	21,032.36	21,728.50	22,633.35	24,080.78
Lubricating oil	7,577.12	7,398.87	9,946.68	7,297.67	6,592.83
Furnace oil	0.00	0.00	0.00	0.00	0.00
Light oil diesel	654.04	11,856.16	436.03	115.42	0.00
Petroleum bitumen	18,120.30	24,496.85	38,147.41	37,916.45	50,729.44
other waste oil	642.69	506.51	15.90	31.77	413.32
Other containing biodiesel	171.62	7.39	0.00	0.00	0.00
Total	571,638.17	569,782.45	605,116.73	635,571.96	693,441.50

Source: Computation Based on IPCC Default Values.

TABLE 4.6

Net Energy Balance of Bhutan from 2013 to 2017, in TJ

Energy Source	2013	2014	2015	2016	2017
Hydro Generation	27,113.22	25,729.56	26,573.76	27,265.82	26,096.11
Fossil fuel	-7,852.76	-7,827.95	-8,297.32	-8,718.05	-9,498.09
Net balance	19,260.46	17,901.61	18,276.44	18,547.77	16,598.02

Source: Computation Based on DGPC and Bhutan Trade Statistics.

to 9,498.09 TJ in 2017 leading to CO_{2e} emission of 571,638.17 tCO_{2e} to 693,441.50 tCO_{2e} between 2013 and 2017. However, with the forest sequestration capacity of 6.3 million tCO_{2e}, the national emission will remain negative (World Bank 2018). This is an important indicator for national carbon neutrality claim. These values are very useful for the net energy balance calculation and associated environmental impact due to energy demand and supply scenario.

Table 4.6 shows the overall energy mix from domestic generation and fossil import, the scenario in which net energy balance between fossil import and electricity generated is on the positive scale dominated by hydro sources, the first indicator for being carbon-negative.

Table 4.7 shows the net energy import and export including both fossil fuel import and the lean season electricity import. Even with the lean season electricity import together with the fossil import, the hydro dominance is by a factor more than 200%. Therefore, the current energy balance shows fossil fuel import dependency has relatively low significance. The grid emission factor currently known is a default value 0.000310319 tCO_{2e}/KWh (UNDP n.d.). Therefore, the current state-of-the art energy profile and environmental impact, the national claim on carbon neutrality is based on hydro as energy source. Owing to fluctuating fossil fuel prices and access limitation, it indicates that there is a potential for a fossil fuel substitute by this clean hydro electrical energy source.

TABLE 4.7

Energy Trading and Net Energy Balance from 2013 to 2017, in TJ

Energy Trade	2013	2014	2015	2016	2017
Generation	27,113.22	25,729.56	26,573.76	27,265.82	26,096.11
Domestic supply	6,846.23	7,431.48	7,711.81	7,504.88	8,342.21
Electricity export	20,333.63	18,645.34	19,107.50	19,742.40	18,246.53
Electricity Import	-389.48	-674.53	-570.46	-398.30	-748.40
Fossil fuel import	-7,852.76	-7,827.95	-8,297.32	-8,718.05	-9,498.09
Net balance	18,937.61	17,574.34	17,951.54	18,130.93	16,342.24

Source: Computation Based on DGPC and Bhutan Trade Statistics.

4.1.2 Scenario 1: Enabling Policy

Policy has remained instrumental and pragmatic in response to sustainable development vision, guided by the visionary and philosophical approach, both in design and implementation under the overarching goal and vision of gross national happiness (GNH), as a baseline development model (Bhutan vision 2020, 1998). Despite limited access to finance, the hydropower development is guided by the sustainable hydropower development policy. Among others, the clean energy development to mitigate global warming has strong policy guide for all hydropower projects (MoEA 2008). The energy policy influences the fossil import and other forms of energy sources. In 2009 Bhutan ranked highest in terms of energy consumption with the 0.87 Ton of Oil Equivalent (TOE/capita) share among the SAARC nations (MoEA 2009). However, electricity dominates the overall energy shares with more than 39%. These figures would have changed by now. The changes will be reflected in scenario 2 of optimization and demand growth. The fossil fuels such as LPG and kerosene are subsidized (MoEA 2009). Despite the fact that major electrical energy source is from hydro, the government has the vision to generate 25 MW of electricity through renewable sources, and new renewable energy policy is put in place for that purpose (MoEA 2013). Additionally, the demand optimization approach through energy efficiency improvement measure in which industrial, household, transport and utility appliances sectors are also adequately addressed in the new energy efficiency policy (DRE 2017). With these four different national policies for energy sector development, namely, national energy policy, sustainable hydropower development policy, alternative energy generation policy and energy efficiency and conservation policy, the national vision for sustainable energy development are inline with the sustainable development goal (SDG) especially in terms of clean energy generation. Thus, the theoretical and visionary baselines are well established for the transition to sustainable energy and energy security questions.

4.1.2.1 Pricing Mechanism

Energy pricing seems to have poor performance in terms of overall electrical energy revenue gain, whereas strong emphasis is made on access and affordability especially for the rural community. The control of national tariffs is based on the regulation set by the Bhutan Electricity Authority (BEA 2017), see Table 4.8.

To ease energy accessibility and affordability, the pricing has been carefully planned, and therefore low-voltage electricity supply receives the highest subsidy ranging from 32% to 100% depending on the final electricity consumed every month. The medium voltage and high voltage are levied for demand charge. Also the low-voltage receives subsidy of up to 28%, whereas the high-voltage power demand does not have any subsidy. While these pricing mechanisms are very favorable for the current situation, industrial development may face serious consequences in the future when the tariff policy is revised. The removal of subsidies will drastically increase the logistic cost for cottage and small-scale industry. Similarly, the changes in the energy subsidy will adversely affect the rural population owing to a low level of income for their livelihood and the limited access to competitive economic growth potential in the future.

TABLE 4.8

National Tariff Plan, A = Generation, B = Network, C = Cost and D = Subsidy are in Nu/kWh E = Final energy tariff after subsidy

Customer Group	DGPC	BPC	Supply	Subsidy	Tariff
	A = Generation	B = Network	C = A + B	D	E = C − D
Low-voltage consumers (230/415 V)					
Block I (Rural Domestic) -<= 100	1.59	4.22	5.81	5.81	Nu.0.00/kWh
Block I (Others) -<= 100 kWh	1.59	4.22	5.81	4.53	Nu.1.28/kWh
Block II - >=101 kWh to <= 300	1.59	4.22	5.81	3.21	Nu.2.60/kWh
Block III ->= 301 kWh	1.59	4.22	5.81	2.38	Nu.3.43/kWh
Low-voltage bulk	1.59	4.22	5.81	1.91	Nu.3.90/kWh
Medium-voltage consumers (6.6/11/33 kV)					
Demand charge					Nu.275/kVA/month
Energy charge	1.59	3.79	5.38	1.53	Nu.2.07/kWh
High-voltage consumers (66 kV and above)					
Demand charge				0	Nu.262/kVA/month
Energy charge	1.59	0.64	2.23	0	Nu.1.59/kWh

Source: BEA 2017.

4.1.2.2 Energy Trading

Energy trading is strongly influenced by the bilateral cooperation between India and Bhutan, referenced to the Jaldhaka Agreement of 1961(MFA 2016). Bhutan received the free power of 250 KW as a royalty payment of Rs.8/KW per annum (Tamang 2007). Although there are many other international support sources for hydropower sector development such as Norway, Japan and other multilateral organizations, the government of India's support has been historic and sustainable. The financing of hydropower infrastructure is either through international grant or loan. For example, the first large-scale electricity generation from Chhukha hydro, the installed capacity of 336 MW, received 60% grant and 40% loan at 5% interest rate from India with the power purchase agreement at Nu.2/kWh(DGPC 2019a). The current hydropower financing modalities have changed from 30– 40% grant and 60–70% loan at 9–10% interest rate on loan (Chophel 2015). Due to these changes, the power purchase agreement also has been slightly changed to 1.80–2.00 Nu./kWh. Owing to the cost escalation for the new hydropower projects that are scheduled to be commissioned, the export prices are expected to drastically increase. For example, the 720 MW Mangdechhu hydro projects were commissioned in 2018 that shows the cost per kWh of electricity is fixed at Nu.4.15 (MoEA 2018). A similar trend will have the consequences on other projects in the pipeline and will adversely affect the domestic tariff which is currently subsidized from 30–100%. On the other hand, Bhutan imports fossil fuel from India where the LPG gas quota of 8,400 MT (5,91,549 cylinders) and 15,000 KL of super kerosene oil per annum at a subsidized rate ranging from 55–60% for LPG and 70–75% for super kerosene fuel (NC 2014). Other fossil fuel prices are affected by fuel price fluctuation in the international market. The energy trade historically has also given a strong foundation of trust and mutual agreement between Bhutan and India.

4.1.3 SCENARIO 2: OPTIMIZATION AND DEMAND PROJECTION

The energy demand and optimization has received quite a lot of attention with different economic and econometric theories owing to the complex nature of planning and energy system design. For example, the World Bank group proposes various models for energy demand and supply scenario (Bhattacharyya and Timilsina 2009). However, the complications involved with the mathematically modeled projection are very incontinent and inconvenient for interpretation when it comes to real world scenario. If at all we can predict the energy future, dependency should be based on industrial growth rate, living conditions and behavior of the consumer. No single modeling method can sufficiently describe the energy demand and supply scenario. Therefore, the projection in this case is purely based on the energy consumption growth rate in various energy demand projections. The growth rates are calculated on an average annual growth rate(AAGR) and compound annual growth rate(CAGR) for a 5-year period starting with the historical data set from (NSB 2018), (NSB 2017), (NSB 2016), (NSB 2015), (NSB 2014) and 2013 as base year in Tables 4.9 and 4.10.

The calculated load demand growth rate on an average is 3%, a deviation of about 30% as compared to the already available data, in which the gross demand growth rate is 4.4% (NTGMP 2018). The growth rate calculations are purely theoretical and based on historical data without any assumption. The projected electrical energy demand for Bhutan is expected to be 6,404.46 GWh with the peak demand of 1,150MW by 2040 at the load demand growth rate of 4.4% (NTGMP 2018). Depending on the demand growth rate, the projected energy demand will also change. Since there is no exact determinant of future growth rate, range of 3–5% may be used for best case and worst case scenario projection. It is quite interesting to know that some electrical energy demand such as industry load and LV load is decreasing. There are two reasons for this, that is either, because of improved efficiency or due to high energy cost industries are closed. Similarly the energy demand for transport sector is expected to increase drastically owing to the projected vehicle growth, which is expected to be 7% in 2040 (DRE 2015). Due to limited access to fuel supply, the transport energy demand will increasingly become challenging in the business as usual scenario. Other energy demand forecast such as fuel wood, heat energy demand and peak load storage are yet to be accessed as their data availability is very limited. Also the heating and cooking is partially combined with electrical load.

The fossil fuel energy demand is in general decreasing such as kerosene, lubricating oil, light oil diesel (LOD), other waste oil and biodiesel. Here we can assume that

TABLE 4.9
Electricity Consumption for 5 Years, in GWh

Type of Consumer	2013	2014	2015	2016	2017	AAGR	CAGR
Domestic rural	73.44	82.09	94.95	103.054	111.841	0.08	0.09
Domestic urban	119.63	127.56	127.9	134.401	137.562	0.03	0.03
Commercial	46.59	51.91	52.35	57.107	63.389	0.06	0.06
Industrial	10.41	8.35	7.89	8.509	8.711	-0.04	-0.04
Agriculture	1.87	2.41	2.36	2.295	2.275	0.03	0.04
Institutions	52.98	55.01	57.65	61.913	59.642	0.02	0.02
Street lighting	3.42	3.96	3.77	3.914	3.983	0.03	0.03
Power house auxiliaries	1.31	1.42	0.91	1.003	0.896	-0.10	-0.07
Temporary connections	22.86	19.2	21.1	23.407	26.274	0.02	0.03
Total LV bulk consumption	69.49	67.94	59.93	64.889	67.055	-0.01	-0.01
Total MV consumption	111.49	89.2	102.15	110.986	114.322	0.00	0.01
Total HV consumption	1,327.97	1,495.78	1,526.17	1437.435	1589.798	0.03	0.04
Total consumption	1,841.46	2,004.83	2,057.14	2008.913	2185.75	0.03	0.03

Source: NSB 2018, Author's Calculation.

the decrease in demand for these fuels is replaced by electricity supply. The only fossil dominance in the future is seen in the transport sector where diesel and petrol are increasing at the average growth rate of 3% and 5%. This growth rate will be applied for future demand projection.

TABLE 4.10
Fossil Fuel Imports in Five Years

Fuel Type	2013	2014	2015	2016	2017	AAGR	CAGR
Diesel	5464.79	5096.29	5465.62	5799.78	6266.41	0.03	0.03
Petrol	1356.19	1409.33	1516.33	1622.42	1752.55	0.05	0.05
Kerosene	222.23	1.07	2.15	1.07	0.54	-41.48	-0.70
Kerosene(S)	2.15	254.40	205.81	215.47	193.68	0.14	1.46
Kerosene (I)	0.00	0.94	0.54	1.07	0.00	0.15	-1.00
Kerosene(J)	141.26	159.86	146.75	146.88	177.56	0.04	0.05
LPG	317.90	333.83	344.88	359.24	382.21	0.04	0.04
Lubricating oil	103.43	100.99	135.77	99.61	89.99	-0.05	-0.03
Furnace oil	0.00	0.00	0.00	0.00	0.00	0.00	-1.00
Light oil diesel	8.84	160.23	5.89	1.56	0.00	-5.60	-1.00
Petroleum bitumen	224.86	303.98	473.38	470.51	629.51	0.17	0.23
Other waste oil	8.77	6.91	0.22	0.43	5.64	-5.94	-0.08
Other containing biodiesel	2.34	0.10	0.00	0.00	0.00	-4.44	-1.00

Source: NSB 2018, Author's Calculation.

TABLE 4.11

Industrial Growth Rate Number of Industries

Sector	2014	2015	2016	2017	2018	AAGR	CAGR
Production & manufacturing	2,823	2,073	1,648	2,125	2,539	−0.23	−0.02
New establishments	335	294	144	445	514	−0.37	0.09
Existing establishments	2,488	1,779	1,504	1,680	2,025	−0.31	−0.04

Source: Author's Calculation.

4.1.3.1 Industrial Growth

Industrial growth in general has been decreasing in Bhutan especially for the small-scale industries. For example, the total industries inclusive of all sectors such as manufacturing and production, service sectors, retail and trade and entertainment in 2018 were 43,924 decreased compared to 70,208 in 2014 (NSB 2018), see Table 4.11.

The new industrial establishment and their growth from the historical data (NSB 2018) show the annual average growth rate is negative, whereas the compound growth rate for the 5-year period is positive. It is also observed from Table 4.12 that small-scale industries' energy demand is decreasing, whereas the large-scale industrial energy demand is increasing. In general, the electrical energy demand is increasing. On the other hand, the fossil energy demand in industrial sector is expected to be decreasing considering the abundant availability of hydro electricity as a clean energy source. In 2042 the expected energy demand is 4,242 GWh as shown in Table 4.12.

However, the decrease in small-scale industries reflects the economic growth is based on large-scale industry such as ferroalloy, cement and mining. Increasing energy demand by promoting cleaner production through development of small-scale industry remains to be low.

4.1.3.2 Livelihood Improvement

Increasing the quality of life in general means the livelihood improvement where various factors influence them, such as better sanitation, good health, good

TABLE 4.12

Industrial Energy Consumption Growth Forecast, in Giga Watt Hour (GWh)

Type of Consumer	CAGR	2017	2022	2027	2032	2037	2042
Industrial	-0.04	8.41	7.03	5.89	4.93	4.12	3.45
Total LV bulk Consumption	-0.01	66.58	64.25	61.99	59.82	57.73	55.70
Total MV consumption	0.01	114.90	117.82	120.81	123.88	127.02	130.23
Total HV consumption	0.04	1,648.06	1,973.00	2,362.00	2,827.70	3,385.22	4,052.67
Total		1,837.94	2,162.09	2,550.69	3,016.33	3,574.09	4,242.07

Source: Author's Calculation.

communication, good transport system, better housing and ease to do manual work. Due to such factors, the living condition in urban areas is to be increasing considering the UN publication on population growth that is 68% of the world population will occupy urban environment by 2050 (UN 2018). In the current situation, Bhutan faces a similar trend. The 2017 population statistics show 274,967 people live in urban areas and 452,178 live in rural areas accounting for less than 40% living in urban areas and more than 60% in rural areas (NSB 2018) at an urban population growth rate of 7.3% (MoWHS 2016). The rural population falls under the 100% energy subsidy scheme for electricity supply where flat peak demand of 2 kilowatt (KW) are planned and the higher consumption are subject to energy tariff subsidy scheme as per the energy tariff policy designed by the Bhutan Electricity Authority (BEA 2017). The energy demand growth is dependent on two factors such as increase in urban consumer and livelihood change in rural areas.

While rural energy demand increases, it also accounts for a shift in use of forest fuel wood of 35,473 m³ (MoAF 2018), annual allotment for fuel consumption and the electrical energy use instead of LPG gas received as annual quota of 8,400 MT (NC 2014) from India. Thus the changes in energy demand will strongly influence the environment. Considering the compound annual growth rate for energy demand, the energy demand is expected to be 3,324.45 GWh in 2042 as shown in Table 4.13.

4.1.3.3   Efficiency Improvement

The energy efficiency improvement can take place through various measures such as behavior change, infrastructure design change and technology innovation. For example, behavior change with the help of Information and Communication Technology (ICT)-enabled smart metering system has the energy-saving potential ranging from 3 to 20% (Simanaviciene, et al. 2014) depending on usage type and scale, infrastructuredesign change can save up to 5–30% of energy demand (Sadineni, Madala and Boehm 2011). Similarly the technology innovation coupled

---

TABLE 4.13
Urban and Rural Energy Demand Forecast, in GWh

Type of Consumer	CAGR	2017	2022	2027	2032	2037	2042
Domestic rural	0.09	111.84	170.32	259.38	395.01	601.55	916.10
Domestic urban	0.03	137.56	209.49	319.03	485.85	739.90	1126.78
Commercial	0.06	63.39	96.53	147.01	223.88	340.95	519.23
Agriculture	0.04	2.28	3.46	5.28	8.04	12.24	18.63
Institutions	0.02	59.64	90.83	138.32	210.65	320.79	488.53
Street lighting	0.03	3.98	6.07	9.24	14.07	21.42	32.63
Power house auxiliaries	-0.07	0.90	1.36	2.08	3.16	4.82	7.34
Temporary connections	0.03	26.27	40.01	60.93	92.80	141.32	215.21
Total		405.86	618.08	941.27	1433.45	2182.99	3324.45

Source: Author's Calculation.

with industrial policy regulation has the potential to save 5–20%(IFC 2015)of final energy consumption.

Although specific energy auditing will give more exact saving potentials, the efficiency improvement assumption from published cases is useful. The energy audit report of Bhutan indicates efficiency improvement potentials through improving load power factors, variable power frequency optimizer, and the energy efficiency improvement for industrial sector ranges between 5 and 40% (DRE 2015) including both electrical and thermal load. The energy demand forecast for best case and worst case scenario can therefore be estimated. Considering the various scopes for efficiency improvement such as behavior change 5–10%, infrastructure design change 6–15% and technology innovation 10–12%, the worst and best case energy demand is 5,405.52 GWh and 5093.78 GWh, respectively, see Table 4.14. The estimate of 21–37% overall efficiency improvement is projected in Table 4.14.

While some fossil fuel demand has been decreasing, such as kerosene, light diesel oil (LOD), waste oil and other bio diesel, few critical fuel demands are still

TABLE 4.14
Energy Improvement and Demand Forecast for Overall Electrical Energy, in GWh

Type of Consumer	Year 2042	Behavior Change		Infrastructure Design		Technology Innovation	
		Worst case (5%)	Best case (10%)	Worst case (6%)	Best case (15%)	Worst case (10%)	Best case (12%)
Domestic rural	916.10	870.30	824.49	818.08	700.82	654.46	616.72
Domestic urban	1126.78	1070.44	1014.10	1006.22	861.99	804.97	758.55
Commercial	519.23	493.26	467.30	463.67	397.21	370.93	349.54
Agriculture	18.63	17.70	16.77	16.64	14.26	13.31	12.54
Institutions	488.53	464.11	439.68	436.26	373.73	349.01	328.88
Street lighting	32.63	30.99	29.36	29.13	24.96	23.31	21.96
Power house auxiliaries	7.34	6.97	6.61	6.55	5.61	5.24	4.94
Temporary connections	215.21	204.45	193.69	192.19	164.64	153.75	144.88
Industrial	3.45	3.28	3.10	3.08	2.64	2.46	2.32
Total LV bulk consumption	55.70	52.92	50.13	49.74	42.61	39.79	37.50
Total MV consumption	130.25	123.74	117.23	116.31	99.64	93.05	87.68
Total HV consumption	4,052.67	3850.04	3647.40	3619.03	3100.29	2895.23	2728.26
Total	7,566.52	7,188.20	6,809.87	6,756.91	5,788.39	5,405.52	5,093.78

Source: Author's Calculation.

TABLE 4.15
Fossil Fuel Import Growth Projection, in TJ

Fuel Type	CAGR	2017	2022	2027	2032	2037	2042
Diesel	0.03	6,266.41	7,185.63	8,239.68	9,448.34	10,834.31	12,423.58
Petrol	0.05	1,752.55	2,009.63	2,304.42	2,642.46	3,030.07	3,474.55
Kerosene(J)	0.05	177.56	203.60	233.47	267.72	306.99	352.02
LPG	0.04	382.21	438.28	502.57	576.29	660.83	757.76
Petroleum bitumen	0.23	629.51	721.85	827.74	949.15	1,088.38	1,248.04
Total		9,208.25	10,558.99	12,107.87	13,883.96	15,920.58	18,255.95

Source: Author's Calculation.

increasing such as diesel, petrol, jet kerosene, LPG and petroleum bitumen as show in Table 4.15. Due to technology maturity and fossil fuel system performance already reaching the maximum possible limit, their efficiency improvement is relatively negligible. Unless transport system is completely switched to e-mobility, the demand for this fossil fuel will keep increasing. At best case efficiency improvement, the net total energy demand in 2042 is expected to be electrical 5,093.78GWh (18,337.61 TJ) and fossil fuel 18,255.95 TJ, for atotal of 36,593.56 TJ.

The national audit report shows the use of coal in large-scale industries such as ferroalloy and cement industries. The imports of all grades of coal, the total coal consumption amounts to 231,558 MT, which constitutes 1,876 MT of anthracite, 98,354.8 MT of sub-bituminous and 90,191 MT of other coal (mostly lignite) and 41,136.7 MT of Coke/Semi Coke (DRE 2015). Further, the rural fuel wood consumption accounts for 35,473 m³(MoAF 2018). Since both fuel sources are expected to be replaced by the electricity sooner or later, their projections are exclusive for net energy balance calculation. However, their significance is visible in environment impact assessment in scenario 3. Even without the consideration of renewable energy sources and their potentials, the net energy balance shows supply surplus domestically.

4.1.4 SCENARIO 3: ALTERNATIVE ENERGY POTENTIALS

The energy sources and national potentials are often time misleading figures. Whereas the energy sources are those resources that can be tapped, the potentials are those resources that are theoretically quantified in terms of net installed capacity. However, the final energy output for useful purposes depends on capacity factors. For Bhutan, the availability of glaciers, annual average solar radiation, available wind speed and the biomass reserve account for renewable energy potentials. The coal deposit is also found in south-eastern parts which account for the fossil fuel resources. The summary of glaciers and glacial lakes are as shown in Table 4.16 (Chhopel et al. 2011). Similarly, the renewable energy resource map shows an average wind speed and daily solar radiation of 4–5 kWh/m² and average wind speed of 4–6 M/S at 50m height from the ground elevation, refer to Figure 4.3 (Gilman, Cowlin and Heimiller 2009).

TABLE 4.16
Summary of Glaciers and Glacial Lakes in Bhutan

Sl	Sub Basins	Glaciers Number	Glaciers Area (Km²)	Glaciers Ice Reserves (Km³)	Glacial Lakes Number	Glacial Lakes Area (Km²)
1	Amo chhu	0	0	0	71	1.83
2	Ha chhu	0	0	0	53	1.83
3	Pa chhu	21	40.51	3.22	94	1.82
4	Thim chhu	15	8.41	0.33	74	2.82
5	Mo chhu	118	169.55	11.34	380	9.78
6	Pho chhu	154	333.56	31.87	549	23.49
7	Dang (Tang) chhu	0	0	0	51	1.81
8	Mangde chhu	140	146.69	11.92	521	17.59
9	Chamkhar chhu	94	104.1	8.11	557	21.03
10	Kuri chhu	51	87.62	6.48	179	11.07
11	Drangme chhu	25	38.54	2.26	126	5.82
12	Nyera Ama chhu	0	0	0	9	0.076
13	Northern basin	59	387.73	51.72	10	7.81
	TOTAL	**677**	**1,316.71**	**127.25**	**2,674**	**106.776**

Source: Chhopel et al. 2011.

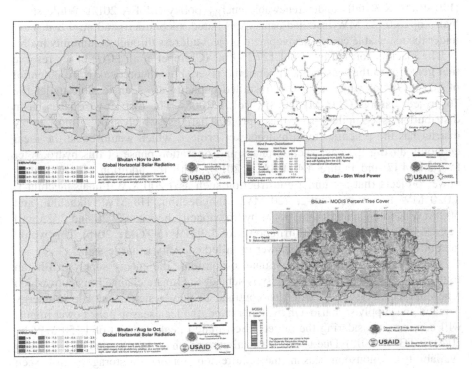

FIGURE 4.3 Energy resource maps of Bhutan (Source: Gilman, Cowlin and Heimiller 2009).

While these resource potentials will remain unchanged, the feasibility to harness them depends on the institutional and financial capacity, which will determine the future energy supply scenario. The biomass potentials are widespread across various sources such as forest residue, agriculture residue, livestock waste and organic waste. The potential analysis for biomass is very dynamic owing to the land use change and population dynamic. While hydropower resources are currently harnessed for both national electricity supply and international export, the solar, wind and biomass are still dormant energy potentials whose exploitation requires further research and detail feasibility study.

The energy generation is mostly dominated by large-scale hydro owing to low cost and good energy market. However, the national demand has indicated the need for backup energy supply in lean season; refer to load profile and peak demand of (BPC 2019) in demand and supply scenario of current situation. Also due to the low renewable energy generation cost, the integration of renewable sources will increasingly become more important. Therefore, the government's plan to add 25 MW (MoEA 2013) of electrical energy installed capacity in the grid will need the renewable energy resource optimization strategy.

4.1.4.1   Hydropower Potential

The sustainable hydropower development policy guides the hydro installed capacity of 1 MW and above. The hydropower is classified as pico, micro, mini, small, and large-scale. The installed capacity of pico (1–10) KW, micro (10–100) KW, and mini (100–1000) KW falls under renewable energy policy (MoEA 2013). While small hydro up to 0–25 MW installed capacity is listed, pico, micro and mini hydro are under the renewable energy development project address through community hydro.

The glaciers formed by about 127.25 km^3 (Chhopel et al. 2011) of ice reserve that give rise to the seasonal and perennial river show theoretical hydro potential of 30.00 GW of which 23.8 GW installed capacity are seen feasible under the 76 project listings (NTGMP 2018) (MoEA 2008). The district-wise river basin distribution is compiled from these project listings, which show almost every district has the potential for a hydropower project, which would boost the local economy during the construction phase and operation phase of this hydro, refer to Table 4.17 and Figure 4.4 (USAID 2019). With the addition of new hydropower projects that are under construction such as 1,200 MW Punatsangchhu-1; 1,020 MW Punatsangchhu-2; 720 MW Mangdechhu; 118 MW Nikachhu; and 600 MW Kholongchhu; the national hydro capacity between 2020 and 2025 will be 5,264 MW with additional firm power of 589MW (NTGMP 2018). This would mean that 22% of total feasible power installed capacity is harnessed.

Therefore, hydropower will continue to dominate the energy supply in Bhutan in the future. However, their capacity is subject to climate change impact to which renewable energy sources will also continue to play a significant role in energy demand and supply scenario. On the other hand, the lean season peak power import will continue, considering the projected estimated peak demand of 1,150 MW by 2040 (NTGMP 2018). Due to the seasonal generation variability and excess power availability in monsoon season, the requirement for energy storage is also seen critical.

TABLE 4.17
Hydropower's Very Rough Overview District-wise

District	River Basin	Number of Hydropower Potentials with Install Capacity (MW)						
		500–2000	100–500	80–100	60–80	40–60	20–40	0–20
Bumthang	Mangdechhu	1	1	0	0	0	0	1
Chhukha	Wangchhu	0	0	0	0	1	0	0
Dagana	Punatsangchhu	0	0	1	0	0	0	1
Gasa	Punatsangchhu	1	0	0	0	0	0	0
Haa	Amochhu	0	0	0	0	1	0	1
Lhuentse	Drangmechhu	0	1	0	0	0	1	0
Monggar	Drangmechhu	1	0	0	0	0	1	0
Paro	Wangchhu	0	1	0	1	0	0	0
Pemagatshel	nil	0	0	0	0	0	0	0
Punakha	Punatsangchhu	2	0	0	0	0	1	0
Samdrup- jongkhar	Others		1	1	1	1	1	0
Samtse	Amochhu	2	0	0	0	3	0	1
Sarpang	nil	0	0	0	0	0	0	0
Thimphu	Wangchhu	0	0	0	0	1	0	0
Trashigang	Drangmechhu	0	0	3	3	1	0	0
Trongsa	Mangdechhu	1	0	0	0	3	0	0
Tsirang	Punatsangchhu	0	3	0	0	0	0	0
Wangdue	Punatsangchhu	0	1	0	2	0	0	0
Yangtse	Drangmechhu	0	2	0	0	0	0	0
Zhemgang	Mangdechhu	3	0	3	0	1	0	0
		11	10	8	7	12	4	4
Total		13,200	3,000	720	490	600	120	40

Source: Author's Compilation.

4.1.4.2 Solar, Wind and Small Hydro

Solar and wind energy potentials depend on the weather conditions and elevation. The first renewable energy resources mapping of 2009 is mostly used today for solar and wind energy potential analysis, see Table 4.18 (Gilman, Cowlin and Heimiller 2009). The maximum wind power density 500 W/m² and maximum wind speed of up to 6 m/s are recorded, which is suitable for electrical energy generation. Similarly average annual solar irradiance is between 4–5 kWh/m², which is suitable for both electricity generation and solar thermal purposes.

Based on the available wind and solar meteorological data, the wind energy generation potential of approximately 761 MW power installed capacity is estimated (DRE 2015). Similarly even if less than 2% of total land area is used for solar PV systems, 9GW power installed capacity is calculated from the annual yield 81,400 GWh from solar. However, the more precise and realistic power capacity is only possible with the site specific design of the detailed project report for renewable power

Figure-2: Basin Maps of Bhutan

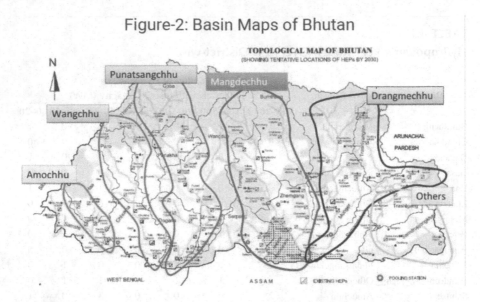

FIGURE 4.4   River basin and hydro potentials maps of Bhutan (Source: USAID 2019).

station development. However, these energy sources have received very insignificant institutional and financial support. As of today, 600 KW wind (Zangpo 2014), 21 KW PV (CST 2018), 11 KW PV (BPC 2018) are seen as an initial phase development after the design of renewable energy policy by DRE (MoEA 2013). Although small-scale solar thermal is also installed in a few office buildings for hot water supply, it is still at the research and development phase. On the other hand, earlier rural house-hold received solar home systems, but they wereremoved in 2015. Similarly, small solar power supplies are also seen to power low-power communication equipment especially the radio and telecommunication power antennas. Therefore, solar and wind energy potential is dormant.

Although the large-scale hydro dominates the electricity generation, small-scale generation and its potentials has been given a strong emphasis. Small-scale hydro-power generation under the renewable energy development projects are those with the installed capacity less than 1 MW. They fall under the pico, micro and mini hydropower. Although many of the already established small-scale projects are de-commissioned, their potentials for the future of renewable energy sources are significant. The estimate of the small-scale power is currently explored by the Department of Renewable Energy for community-based hydro electricity supply. There are 17 different known small hydros with the installed power capacity ranging from 10 to 200 KW in various communities in different districts, see Table 4.18. Total power installed capacity of these small-scale generations is 1,480 KW (Dorji 2007).

Considering the current state-of-the-art technology and the strong focus on large-scale hydropower project, the small-scale generation will continue to be neglected. Whereas wind and small-scale hydro is suitable for electricity generation, solar potential offers various options such as solar thermal for household heat energy demand requirement, outdoor solar cooking stove for tourism, agriculture use for

TABLE 4.18
Solar and Wind Potentials

District	Elevation	Wind Density	Speed (m/s)	Area	Solar Irradiance	DC	AC	Hydro Capacity
Bumthang	1800–5400	400–500	5.5–5.9	40.8	5.2	6,700	6,000	80
Chhukha	600–4800	300–400	5.0–5.5	28.2	4.8	4,300	3,800	0
Dagana	600–4800	200–300	4.4–5.0	25.9	4.8	3,900	3,500	200
Gasa	1800–5400	300–400	5.0–5.5	47.0	5.3	7,800	7,000	0
Haa	1800–5400	300–400	5.0–5.5	28.6	5.0	4,500	4,000	0
Lhuntse	1800–5400	400–500	5.5–5.9	42.9	4.9	6,600	5,900	240
Mongar	600–4200	300–400	5.0–5.5	29.2	4.8	4,400	4,000	200
Paro	1800–5400	300–400	5.0–5.5	19.3	5.3	3,200	2,900	0
Pemagatshel	600–3000	300–400	5.0–5.5	15.3	5.0	2,400	2,200	0
Punakha	1200–5400	400–500	5.5–5.9	16.7	5.0	2,600	2,400	0
Samdrup Jongkhar	200–4200	0–200	0.0–4.4	28.2	4.7	4,100	3,700	0
Samtse	200–4800	0–200	0.0–4.4	19.6	4.7	2,900	2,600	0
Sarpang	200–4200	0–200	0.0–4.4	24.8	4.7	3,700	3,300	70
Thimphu	1800–5400	200–300	4.4–5.0	26.9	5.3	4,500	4,000	40
Trashigang	600–4800	300–400	5.0–5.5	33.1	4.9	5,100	4,600	0
Trongsa	600–5400	300–400	5.0–5.5	27.2	4.9	4,200	3,700	190
Tsirang	600–4200	400–500	5.5–5.9	9.57	4.9	1,500	1,300	200
Wangdue Phodrang	600–5400	400–500	5.5–5.9	60.6	5.1	9,600	8,600	40
Trashiyangtse	600–5400	400–500	5.5–5.9	21.7	4.9	3,400	3,000	0
Zhemgang	600–4200	300–400	5.0–5.5	36.3	4.8	5,500	4,900	220

Source: Gilman, Cowlin and Heimiller 2009; Dorji 2007.

solar water pumps and solar fencing. Therefore, even if direct electricity generation is not competitive compared to the large-scale hydro at the moment, the renewable energy diversification to alternative use will increasingly supplement the national energy demand requirement. However, the alternative energy source identified will continue to support as an alternative to climate change adaptation measure in the worst case scenario.

4.1.4.3 Biomass Potentials

The biomass potential depends on the availability of agriculture residue, forest residue, organic waste, waste water and livestock waste. Since Bhutan is not rich in agriculture, the agriculture waste is mostly consumed by livestock. To compile the forest residue for biomass potential is difficult, as data is not available. Therefore, the biomass potentials are calculated from the approximate per capita household waste 0.235 kg (DRE 2015) and an assumption that one cattle produces 1.5 kg of dung per day.

Biomass will be the source of energy supply for thermal purposes in many developing countries including Bhutan. Whereas fuel wood as biomass resources for thermal use is perceived as unsustainable, the development of biogas technology offers range of fuel wood substitutes. Biogas is a relatively old technology, whereas the biogas from waste has emerged as an innovation in energy generation from waste. For potential analysis, the household waste generates 102 m^3/t of CH$_4$ gas (IEA 2014) and 0.023–0.040 m^3/kg biogas from cattle dung (FAO 1996). The energy content of CH$_4$ is 36MJ/m^3 or 10 kWh and that of biogas 18–27 MJ/m^3 or 5.0–7.5 kWh/m^3 (FNR 2017). The combined energy potential of 1,117.54 GJ is calculated from the available feedstock for small-scale biogas plant, see Table 4.19 population data (NSB 2017). The increasing fuel demand, especially in rural areas, has been reduced by the adoption of the available biogas potentials and through rural electrification, refers to the fuelwood demand in Table 4.20. Since fuel woods used are not easily measurable owing to theircomplexity in approximation,

TABLE 4.19
Biomass Potential from Household Waste and Cattle Dung

District	Population	Human Waste	CH$_4$	Population	Animal Dung	Biogas	Energy
Bumthang	17,820	4,188	427.15	11,311	16,967	678.66	33.70
Chhukha	68,966	16,207	1,653.12	19,020	28,530	1,141.20	90.32
Dagana	24,965	5,867	598.41	18,853	28,280	1,131.18	52.08
Gasa	3,952	929	94.73	1,163	1,745	69.78	5.29
Haa	13,655	3,209	327.31	9,032	13,548	541.92	26.42
Lhuentse	14,437	3,393	346.05	13,468	20,202	808.08	34.28
Monggar	37,150	8,730	890.49	26,782	40,173	1,606.92	75.44
Paro	46,316	10,884	1,110.19	14,119	21,179	847.14	62.84
Pemagatshel	23,632	5,554	566.46	8,872	13,308	532.32	34.77
Punakha	28,740	6,754	688.90	11,045	16,568	662.70	42.69
Samdrup Jongkhar	35,079	8,244	840.84	13,807	20,711	828.42	52.64
Samtse	62,590	14,709	1,500.28	31,457	47,186	1,887.42	104.97
Sarpang	46,004	10,811	1,102.72	22,850	34,275	1,371.00	76.71
Thimphu	138,736	32,603	3,325.50	5,011	7,517	300.66	127.84
Trashigang	45,518	10,697	1,091.07	25,730	38,595	1,543.80	80.96
Trashi Yangtse	17,300	4,066	414.68	11,661	17,492	699.66	33.82
Trongsa	19,960	4,691	478.44	11,583	17,375	694.98	35.99
Tsirang	22,376	5,258	536.35	11,335	17,003	680.10	37.67
Wangdue Phodrang	42,186	9,914	1,011.20	23,306	34,959	1,398.36	74.16
Zhemgang	17,763	4,174	425.78	12,110	18,165	726.60	34.95
Total	**727,145**	**170,879**	**17,430**	**302,515**	**453,773**	**18,151**	**1,117.54**

Source: Author's Analysis from NSB 2017.

TABLE 4.20
Fuel Wood Used in m³

	2013	2014	2015	2016	2017
Fire wood	32,866.91	35,988.34	40,490.66	38,184.81	34,452.00
Wood chips	12,517.07	10,297.18	8,004.91	2,250.68	6,621.76
Briquettes	0.00	0.00	0	266,580.00	185,340.00

Source: NSB 2018.

the data in Table 4.20 is used only to give a rough idea which is accumulated from the national statistics (NSB 2018). More research and field data measurement is required for this. Similarly the biogas potentials in Table 4.19areonly from the cattle, and other livestock are excluded as they represent relatively low percentile of waste and often the animal housings are together. However, the potentials from household waste and cattle dung are an option for fuel wood substitute, where the forest protection and clean fuel source are added value gained from such an approach.

The biogas as a rural cooking fuel has become an attractive alternative for rural community, where livestock are kept indoors and the idea of circular economy as an approach to rural economy is considered to be successful. As of 2016, there are 3,044 biogas plants with the size range 4–10 m³ are already constructed in different communities (SNV 2016). The medium size biogas plants that range from 70 to 100 m³ size are also feasible, where the major livestock farms are available. Similarly, the biogas potentials from household waste are very recently seen as an opportunity to solve organic waste challenges in urban centers, boarding schools, institutes and colleges. However, their potentials are still not utilized owing to financial and institutional capacity to implement them.

4.1.4.4 Levelized Cost of Renewable Energy

The Levelized Cost of Electrical (LCOE) energy is the cost of generation expressed in $/kWh. The cost of generation is influenced by the financial regulation such as tax and subsidy, logistic cost, operation and maintenance cost, interest rate, net annual generation and total operating hours of the power plants. Since dependency of LCOE is plant specific and has energy generation potential, its exact calculations is normally based on the already constructed plant and during the detailed project report preparation. However, the installation of cost per KW_p can be compared with the existing system. The average market price of renewable energy technology per KW_p and LCOE is as shown in Table 4.21.

The average sunshine hour in Bhutan is 5–8 hr/day on average (Climate – Bhutan, n.d.). The solar capacity factor of 30% and wind capacity factor of 25% (Kuensel 2018) are normally possible as seen in the general weather forecast. Similarly, the hydro has average capacity factor of 50% and biogas capacity factor range from 70 to 90%. Comparing the current large hydro export price of Nu.4.15/kWh (MoEA 2018) with the renewable sources, they are already able to compete, see Table 4.21. When such

TABLE 4.21

Cost Per KW Peak and LCOE

	PV$_{Small}$	PV$_{Medium}$	PV$_{Large}$	Wind$_{onshore}$	Wind$_{offshore}$	Biogas$_{Waste(H)}$	Hydro $_{micro}$
Low (€)	1,560	1,040	780	1,950	4,030	2,600	1,000
High(€)	1,820	1,300	1,040	2,600	6,110	5,200	2,000
LCOE(€)	(7-12)€$_{ct}$	(5-8)€$_{ct}$	(3-5)€$_{ct}$	(4-8)€$_{ct}$	(5-9)€$_{ct}$	(10-15)€$_{ct}$	(1.4-6.5)€$_{ct}$
LCOE($)	(9-16)$$_{ct}$	(7-10)$$_{ct}$	(4-7)$$_{ct}$	(5-10)$$_{ct}$	(7-11)$$_{ct}$	(13-19)$$_{ct}$	(2-8.45)$$_{ct}$
LCOE (Nu.)	(7-12.48)	(5-8)	(3-5.5)	(3.9-7.8)	(5.5-8.58)	(10-15)	(1.56-6.6)

Source: Kost et al. 2018; IRENA 2012.

costs margin is achieved, the large hydro will have to face the challenge of declining renewable energy cost; therefore, it strongly affects the international energy trading. On the other hand, with the removal of the current subsidy for domestic electricity tariff, the actual cost of electricity will be 5.58 Nu/kWh (BEA 2017), which is slightly higher than the solar PV LCOE; therefore, it also strongly supports the feed-in tariff policy. Considering the declining cost of generation from renewable energy shows that the renewable vision of 25 MW (MoEA 2013) generation by DRE is achievable.

4.1.4.5   Fossil Deposit in Bhutan

SD Eastern Bhutan Coal Company Limited (SDEBCCL) is a privately owned company which started coal mining in Bhutan in the early 1990s in the south-eastern parts of Bhutan, in the Samdrup Jongkhar district. Since the data availability and scientific records are very limited, the only fossil mining known in Bhutan is coal. The exact information on coal is also not clearly known in Bhutan. For example, currently 600,000MT (G. K. Dorji 2016) of coal deposit is predicted in Habrang coal mine, which was abandoned by SDEBCCL involved earlier in mining, considering the deposit is fully exploited. The coal energy is used for thermal purposes in ferroalloy industries domestically. Since the exact availability is unknown, and also the push toward the use of electricity is foreseen in energy demand, the coal will potentially end when other renewable energy dominates the thermal energy supply.

Tables 4.22 and 4.23 show the total coal used domestically and exported both in metric ton (MT) units and energy units, terajoule (TJ). Small fractions of forest

TABLE 4.22

Coal Use in Metric Tons (MT)

Coal	2013	2014	2015	2016	2017
Domestic Use	77,743.58	95,317	79,762	117,783	142,596
Export	0	26,574	5,403	0	18,931
Total	77,743.58	121,891	85,164	117,783	161,527

Source: NSB 2018.

TABLE 4.23
Energy Content in Terajoule (TJ)

Coal	2013	2014	2015	2016	2017
Domestic Use	2,278.66	2,794	2,338	3,452	4,179
Export	0	779	158	0	555
Total	2,278.66	3,572.63	2,496.16	3,452.22	4,734.36

Source: Author's Analysis.

wood were also used for thermal energy demand by the localized paralysis process by small private cooperatives wherever the industrial coal demands are required. The feasibility study is done for thermal energy demand fulfillment through the use of bamboo as feedstock (MoEA 2015). Therefore, the use of coal for thermal demand will be substituted by electrical and biomass use in future. Therefore, coal as energy potential is rather viewed as mining industry instead of energy sources.

4.2   ENVIRONMENTAL IMPACT

Perhaps a strong belief system and good governance have much greater influence on environmental protection than the modern environmental activist for the environment studentshipin Bhutan (Kaewkhunok 2018). So for the first phase of Bhutan's economic transition and exposure to a modern energy system, there is no negative effect on the environment except for the shifting cultivation practice, which the government abolished with the deployment of the Forest Act in 1969(FAO n.d.). However, the economic growth coupled with development of large-scale hydropower shows that the per capita CO_{2e} emission has increased from 0.01 metric ton in 1970 to 1.29 metric ton in 2014(World Bank 2014). This data needs verification as for the energy sector development and associated economic growth; grid emission factor is critical for CO_{2e} emission calculation, which has received less attention. More detail on emission calculation is shown in Table 4.25 as well as the impact of energy generation and consumption on the environment.

The main source of environmental pollution is from the transport sector. The emission data are approximate, which is mostly secondary information. 261,887 tons of CO_{2e}, 1,724 tons of NO_2, 38.9 of particulate matter (PM) counts and 45 SO_2 contains are recorded as of 2015(ADB 2019). The transport sector is already in a transition phase with the policy revision in 2016. The new transport policy addressing inclusivity for various options identifies e-mobility as a sustainable and resilient mobility mode in the future.

The impact of energy demand and supply influences three fundamental aspects such as human health, environment health, and biodiversity (Kjellstrom et al. 2000). They are the global indicators and known since the Industrial Revolution. However, their significance isclearly revealed and emphasized after the publication of the Club of Rome report on limits to growth model (Meadowset al. 1972). The summary of energy consumption, generation, import and export is as shown in Table 4.24 in terajoule (TJ) units.

TABLE 4.24

Summary of Energy Use by Type in Bhutan, in TJ

Energy Source	2013	2014	2015	2016	2017
		Electrical			
Domestic	6,846.23	7,431.48	7,711.81	7,504.88	8,342.21
Import	389.48	674.53	570.46	398.30	748.40
Total	7,235.71	8,106.01	8,282.27	7,903.19	9,090.61
		Mobility/Transport			
Diesel	5,464.79	5,096.29	5,465.62	5,799.78	6,266.41
Petrol	1,356.19	1,409.33	1,516.33	1,622.42	1,752.55
Kerosene(J)	141.26	159.86	146.75	146.88	177.56
Total	6,962.25	6,665.48	7,128.69	7,569.08	8,196.53
		Thermal			
Coal	2,278.66	2,793.74	2,337.82	3,452.22	4,179.49
Petroleum bitumen	224.86	303.98	473.38	470.51	629.51
LPG	317.90	333.83	344.88	359.24	382.21
Fire wood	153.82	168.43	189.50	178.70	161.24
Wood chips	58.58	48.19	37.46	10.53	30.99
Briquettes	0.00	0.00	0.00	2,963.04	2,060.05
Total	3,033.82	3,648.17	3,383.04	7,434.24	7,443.49
		Others/Miscellaneous			
Kerosene	222.23	1.07	2.15	1.07	0.54
Kerosene(S)	2.15	254.40	205.81	215.47	193.68
kerosene (I)	0.00	0.94	0.54	1.07	0.00
LOD	8.84	160.23	5.89	1.56	0.00
Other waste oil	8.77	6.91	0.22	0.43	5.64
Other containing biodiesel	2.34	0.10	0.00	0.00	0.00
Lubricating oil	103.43	100.99	135.77	99.61	89.99
Total	**347.76**	**524.66**	**350.37**	**319.23**	**289.85**

Source: Author's Analysis.

The development model of Bhutan has always been based on the careful planning approach which can be seen clearly today in GNH-based policy screening criteria (Ura et al. 2012), where the major focus is on environmental conservation and sustainable socio-economic growth. The energy generation and consumption policies based on the GNH policy have strong influence on various energy policies, such as sustainable hydropower (MoEA 2008), alternative renewable energy (MoEA 2013), and energy efficiency policy (DRE 2015), which have a strong influence on environmental conservation and the national commitment to remain carbon-neutral. The de-carbonization approach for sustainable socio-economic growth has strong focus on forest conservation. With the forest coverage of more than 70%, the carbon sequestration capacity of 6.3 million tCO2e (World Bank 2018) is identified as a significant contribution at the moment in the SAARC region with respect to climate

TABLE 4.25
tCO₂ₑ Emission in 5 Years

Energy Source	EF tCO₂ₑ/TJ	2013	2014	2015	2016	2017	Remarks
Electricity	0	0	0	0	0	0	GEF still developing
Electrical							
Mobility/Transport							
Diesel	74.74	408,438.57	380,896.91	408,500.45	433,475.22	468,351.85	IPCC default value
Petrol	69.93	94,838.69	98,554.26	106,036.79	113,456.04	122,556.07	IPCC default value
Kerosene(J)	72.15	10,191.99	11,533.55	10,587.70	10,597.35	12,810.76	IPCC default value
Total		513,469.24	490,984.72	525,124.95	557,528.62	603,718.69	
Thermal							
Coal	94.6	215,561.65	264,287.92	221,158.17	326,579.99	395,379.64	IPCC default value
Bitumen	81.4	18,303.33	24,744.29	38,532.73	38,299.45	51,241.86	IPCC default value
LPG	63.1	20,059.40	21,064.54	21,761.75	22,667.98	24,117.63	IPCC default value
Fire wood	109.6	16,858.36	18,459.43	20,768.79	19,586.06	17,671.40	IPCC default value
Wood chips	109.7	6,426.21	5,286.53	4,109.69	1,155.49	3,399.59	IPCC default value
Briquettes	109.8	0.00	0.00	0.00	325,341.43	226,193.94	IPCC default value
Total		277,208.95	333,842.71	306,331.13	733,630.39	718,004.04	
Others/Miscellaneous							
Kerosene	72.52	16,116.04	77.89	155.77	77.92	38.94	IPCC default value
Kerosene(S)	72.52	155.77	18,449.36	14,925.00	15,625.98	14,045.53	IPCC default value
kerosene (I)	72.52	0.00	68.15	38.94	77.89	0.00	IPCC default value
LOD	74.74	660.65	11,975.91	440.43	116.59	0.00	IPCC default value
Waste oil	74.00	649.18	511.63	16.06	32.09	417.50	IPCC default value
Biodiesel	74.00	173.36	7.46	0.00	0.00	0.00	IPCC default value
Lubricating oil	74.00	7,653.65	7,473.60	10,047.15	7,371.39	6,659.43	IPCC default value
Total		25,408.66	38,564.01	25,623.35	23,301.84	21,161.40	
Gross total		816,086.85	863,391.44	857,079.44	1,314,460.85	1,342,884.12	

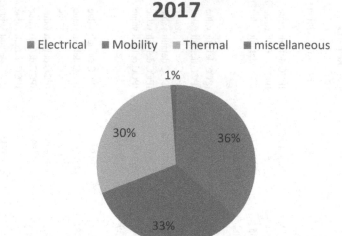

FIGURE 4.5   Energy consumption by type in % share (Source: author's analysis).

commitment. The summary of emission and overall carbon balance is shown in Table 4.25. Owing to various industrial development and economic growth, emission is unavoidable and impossible to remain in an ideal case; however, a clean energy source has the highest contribution to environmental protection and sustainable socio-economic growth in Bhutan.

Table 4.24 shows the summary of overall energy consumption in Bhutan and Figure 4.5 shows the percentile share of energy mix for 2017. The energy mix is dominated by electricity with 36% share followed by the mobility sector with 33% and thermal energy demand of 30%. The miscellaneous has 1% share which is infact decreasing. Both thermal and miscellaneous aredecreasing owing to 100% electrification and use of electricity for cooking and heating purposes. Since the hydro electricity supply is considered to be a clean source compared to fossil fuel, the future energy mix is expected to be covered by the electricity supply. In the current scenario, total energy consumption has increased from 17,579.54 TJ in 2013, to 25,020.47 TJ in 2017. With an increasing share of clean source in energy mix, energy has contributed an environment to resiliency, see Table 4.24. The approximation error of 10–20% should be used for the energy, as data inconsistency was noted in different literature sources. Also the fuel wood data are mostly approximate figures, as their measurements are not exact, owing to the complexity of weighting them and identifying their moisture content.

Table 4.25 shows the summary result of the overall energy and environment of Bhutan. Since the grid emission factor (GEF) 0.000310319000 tCO_{2e}/kWh (UNDP n.d.) is a default value, which is currently under revision, the emission from electricity generation is considered to be zero in Table 4.25. However, the grid emission will have to be considered for a more detailed analysis. The net emission due to fossil fuel consumption has increased from 816,086.85 tCO_{2e} in 2013 to 1,342,884.12 tCO_{2e} in 2017, see Table 4.25. However with the current forest sequestration capacity of 6,300,000.00 tCO_{2e}/kWh, the national emission remains negative. For the grid emission factor of 0.000310319000 t CO_{2e}/kWh, the CO_{2e} emission for 2017 is 2,249,477.61 tCO_{2e} for electricity generation from hydro and the total emission due to energy consumption and generation is 3,592,361.73 tCO_{2e}, which is 1.75 times less than that of forest sequestration capacity. Thus, Bhutan has a good reason to be carbon-negative in the SAARC region.

4.3   CONCLUSION

Reflecting on the past and looking into the future, Bhutan's energy scenario has meaningful insight to the global energy challenges. In the current scenario, 100% access to electricity in 2015 within the span of approximately 30 years, Bhutan started exploring modern energy systems, and economic growth is amazingly an achievement to be shared internationally. Additionally, the energy source is hydro based, where the emission is relatively low compared to fossil and nuclear energy sources. This also showcases 100% renewable energy vision as a feasible option for sustainable socio-economic growth and environmental studentship. The projected energy demand of 36,593.56 TJ in 2042 is expected to be supplied through a different renewable mix such as hydro, solar, wind and biomass, shifting the fossil era to a sustainable energy future. However, thisissubject to various challenges.

Despite numerous challenges, which are mostly the need of the orientation(proper management), the energy profile of Bhutan has a strong message to convey in the region, philosophically and strategically. The vision of peace, prosperity and happiness is core to human development that the country has invested in as of now. Robust and careful planning, along with the alignment of its policy and practices to best international practices, is amajor contribution to energy and the environment.

Whereas the rough estimates and management approaches are covered, the limitation in this chapter is lack of primary data, as it is practically not feasible to complete the detailed onsite investigation in this short time span. However, the author has invested a lot of time digging out the national statistics, and calculative approach is used to compare the available report and literature sources. Therefore, less descriptive but more statistical information are provided for the reader in this chapter of Bhutan energy and environment. It is recommended that there are good energy research potentials in Bhutan, especially in areas of thermal energy demand and mobility energy demand. Energy auditing for the future researcher is valuable energy research work and will increasingly become a national interest for energy data development for the national commitment to remain carbon-neutral.

REFERENCES

ADB. Bhutan Vehicle Emission Reduction Road Map and Strategy, 2017–2025. Thimphu. Asian Development Bank, online @https://www.adb.org/sites/default/files/publication/513931/adb-brief-110-bhutan-vehicle-emission-reduction-strategy.pdf, 2019.

Barsky, Robert, B. and Lutz Kilian. "Oil and the Macroeconomy since the 1970s." *Journal of Economic Perspectives*, online @https://pdfs.semanticscholar.org/a23d/7f31367c52a13 37a563c64298760add31d29.pdf, 2004: 1–18.

Baruah, Ditee Moni. "The Refinery Movement in Assam." *Economic and Political Weekly*, online @ https://www.jstor.org/stable/27917990?seq=1#page_scan_tab_contents, 2011: 1.

BEA. Electricity Tariff in Bhutan.Tariff Plan Report, Thimphu: Bhutan Electricity Authority, online @ http://www.bea.gov.bt/wp-content/uploads/2017/12/Electricity_Tariff_in_ Bhutan_2017.pdf, 2017.

Bhattacharyya, Subhes C., and Govinda R.Timilsina. Energy Demand Models for Policy Formulation: A Comparative Study of Energy Demand Models. Research finding, World Bank, online @ http://documents.worldbank.org/curated/en/800131468337793239/pdf/ WPS4866.pdf, 2009.

Bhutan vision2020.A Vision for Peace, Prosperity and Happiness. Thimphu: Planning Commission Royal Government of Bhutan, online @ http://unpan1.un.org/intradoc/ groups/public/documents/APCITY/UNPAN005249.pdf, 1998.

BPC. Annual performance report. Corporate performance report, Thimphu: Bhutan Power Corporation, online @ http://bpso.bpc.bt/wp-content/uploads/2018/02/Annual-Report-2017.pdf, 2017.

BPC. Annual power system performance report.Corporate performance report, Thimphu: Bhutan Power Corporation, online @ http://bpso.bpc.bt/wp-content/uploads/2017/07/ Annual-Report-2016.pdf, 2016.

BPC. Annual Report. Corporate report public information, Thimphu: Bhutan Power Corporation, online @ https://www.bpc.bt/wp-content/uploads/2019/05/BPC-Annual-Report-2018.pdf, 2018.

BPC. Annual report power system performance. Corporate performance report, Thimphu: Bhutan Power Corporation, online @ http://bpso.bpc.bt/wp-content/uploads/2019/01/ Annual-Report-2018.pdf, 2018.

BPC. Transmission System Performance Quarterly Report. Performance indicators, Thimphu: Bhutan Power Corporation, online @ http://bpso.bpc.bt/wp-content/uploads/2019/07/ Second-Quartely-Report-2019.pdf, 2019.

Chhopel, G. Karma, et al. Securing the Natural FreshwaterSystems of the Bhutan Himalayas. Project report, online @ http://sa.indiaenvironmentportal.org.in/files/file/Water_ Paper_Bhutan.pdf, 2011.

Chophel, Sangay. "Export Price of Electricity in Bhutan: The Case of Mangdechhu Hydroelectric Project." *Journal of Bhutan Studies*, Vol.32: 1–5. Online @http://www. bhutanstudies.org.bt/publicationFiles/JBS/JBS_Vol32/1.%20Export%20price%20 of%20electricity.pdf, 2015.

DGPC. Annual Report.Corporate report for public, Thimphu: Druk Green Power Corporation, online @ http://www.drukgreen.bt/wp-content/uploads/2019/06/Annual-Report-2018-1.pdf, 2018.

DGPC. Chhukha Hydropower Plant.Chhukha: Druk Green Power Corporation,online @ http:// www.drukgreen.bt/wp-content/uploads/2019/01/Chhukha-Hydropower-Plant.pdf, 2019a.

DGPC. Druk Green Power Corporation. http://www.drukgreen.bt/about-us/, 2019b.

Dorji, Gyalsten K. "Habrang coal mine leased to SMC." *Kuensel*, 1–2, online @ http://www .kuenselonline.com/habrang-coal-mine-leased-to-smc/,2016.

Dorji, Karma Penjor. The Sustainable Management of Micro Hydropower Systems for Rural Electrification: The Case of Bhutan. *Master thesis*, online @ http://humboldt-dspace .calstate.edu/bitstream/handle/2148/287/k.dorji.pdf?sequence=1, 2007.

DRE. Bhutan Energy Data Directory.Thimphu: Department of Renewable Energy, 2015.

DRE. Consolidated Energy Audit Report for Industries. *Consultancy energy audit report, Thimphu: Department of Renewable Energy*, online @ http://www.moea.gov.bt/wp-content/uploads/2018/07/Consolidated-Industry-Audit-report-Final.pdf, 2015.

DRE. Energy Efficiency in Transport Sector. Consultancy report, Thimphu: Department of Renewable Energy, online @ http://www.moea.gov.bt/wp-content/uploads/2018/07/ Final-Transport-EE-report.pdf, 2015.

DRE. National Energy Efficiency & Conservation Policy. Policy draft, Thimphu: Department of Renewable Energy, online @ https://www.gnhc.gov.bt/en/wp-content/uploads/2017/05/EEC-Final-Draft-Policy-2017-Final-1.pdf, 2017.

FAO. http://www.fao.org/3/v8380e/V8380E05.htm (accessed 2019).

FAO. Support for Development of National Biogas Programme. Biogas Technical Manual, Kathmandu: United Nations Nepal, online @ https://sswm.info/sites/default/files/reference_attachments/FAO%201996%20Biogas%20Technology%20A%20Training%20Manual%20for%20Extension.pdf, 1996.

FNR. Bioenergy in Germany: Facts and Figures.Facts and Figures International, Germany, online @ http://www.fnr.de/fileadmin/allgemein/pdf/broschueren/broschuere_basis-daten_bioenergie_2017_engl_web.pdf, 2017.

Gilman, Paul, Shannon Cowlin, and Donna Heimiller. Potential for Development of Solar and Wind Resource in Bhutan. Consultancy Report, USA: National Renewable Energy Laboratory, online@ https://www.nrel.gov/docs/fy09osti/46547.pdf, 2009.

Harris, Frances.Gross National Happiness for Children: Embedding GNH Values in Education. Thimphu: Centre for Bhutan Studies and Gross National Happiness, 2017.

IEA. Waste to Energy. Summary fact sheet, Austria: UNFCCC, online @ https://unfccc.int/sites/default/files/resource/Waste%20to%20energy%202.pdf, 2014.

IFC. Study on Energy Efficiency and Energy Saving Potential in Industry and on Possible Policy Mechanisms. EU consultancy report, European Commission Directorate-General Energy,online @ https://ec.europa.eu/energy/sites/ener/files/documents/151201%20DG%20ENER%20Industrial%20EE%20study%20-%20final%20report_clean_stc.pdf, 2015.

IPCC. Guidelines for National Greenhouse Gas Inventories.Module 1:Energy.IPCC, online @ https://www.ipcc-nggip.iges.or.jp/public/gl/guidelin/ch1wb1.pdf, 1996.

IRENA. Renewable Energy Technologies: Cost Analysis Series. Public information on LCOE, International Renewable Energy, online @ https://www.irena.org/document-downloads/publications/re_technologies_cost_analysis-hydropower.pdf, 2012.

Kaewkhunok, Suppawit. Environmental Conservation in Bhutan: Organization and Policy.Institute of Asian Studies, Chulalongkorn University, 1–10, 2018.

Kjellstrom, Tord et al. "Energy, the Environment, and Health." In World Energy Assessment: Energy and the Challenge of Sustainability, by John P.Holdren and Kirk R.Smith, 1–5, online @ http://user.iiasa.ac.at/~gruebler/Lectures/Graz-04/wea_chapter3.pdf, 2000.

Kocaslan, Gelengul. "International Energy Security Indicators and Turkey's Energy Security Risk Score." International Journal of Energy Economics and Policy, online @ http://citeseerx.ist.psu.edu/viewdoc/download?doi=10.1.1.1016.180&rep=rep1&type=pdf, 2014: 1–3.

Kost, Christoph, Thomas Schlegl, Hans-Martin Henning, and Andreas Bett. Levelized Cost of Electricity: Renewable Energy Technologies. Research report, Germany: Fraunhofer Institute for Solar Energy Systems Ise, online @https://www.ise.fraunhofer.de/content/dam/ise/en/documents/publications/studies/EN2018_Fraunhofer-ISE_LCOE_Renewable_Energy_Technologies.pdf, 2018.

Kuensel. "Wangdue windmills generate about 2M units of electricity." Kuensel online @ http://www.kuenselonline.com/wangdue-windmills-generate-about-2m-units-of-electricity/, May 2018: 1-2.

Meadows, Donella H., Dennis L.Meadows, JórgenRanders, and William W.Behrens. "The Limits to Growth." In The Limits to Growth, 1–10. New York: Club of Rome, online @ http://donellameadows.org/wp-content/userfiles/Limits-to-Growth-digital-scan-version.pdf, 1972.

MFA. Royal Bhutan Embassy, New Delhi India. 2016. https://www.mfa.gov.bt/rbedelhi/?page_id=28.MoAF. Forest Facts & Figures.Thimphu: Department of Forests and Park Services, online @http://www.dofps.gov.bt/wp-content/uploads/2016/03/FFF-2018.pdf, 2018.

MoEA. Alternative Renewable Energy Policy. RE policy, Thimphu: Department of Renewable Energy, online @ http://img.teebweb.org/wp-content/uploads/2015/06/alternative-renewable-energy-policy-2013.pdf, 2013.

MoEA. Bhutan energy policy. National energy policy, Thimphu: Ministry of Economic Affairs, online @ https://eneken.ieej.or.jp/data/2598.pdf, 2009.

MoEA. Bhutan sustainable hydropower development policy. Government policy, Thimphu: Ministry of Economic Affairs, online @ http://www.moea.gov.bt/wp-content/uploads/2017/07/Hydropower-Policy.pdf, 2008.

MoEA. Detailed Feasibility Report: Bamboo Charcoal Manufacturing Unit. Feasibility study, Thimphu: Ministry of Economic Affairs, online @ http://www.moea.gov.bt/wp-content/uploads/2017/07/Bamboo-charcoal.pdf, 2015.

MoEA. MoEA Newsletter,1–10, online @ http://www.moea.gov.bt/wp-content/uploads/2019/08/MoEA-Newsletter-Volume-I-Issue-I.pdf,2018.

MoWHS. Infrastructure Development in Bhutan: A journey through time.Thimphu: Ministry of Works and Human Settlement, online @ https://www.mowhs.gov.bt/wp-content/uploads/2014/03/Final_MoWHS_Magazine_2015-1.pdf, 2015.

MoWHS. The 3rd UN Conference on Housing and Sustainable Urban Development. UN conference report, Thimphu: Ministry of Works and Human Settlement, online @ http://habitat3.org/wp-content/uploads/Bhutan_Habitat-III-National-Report.pdf, 2016.

NC. A Review of Issues Pertaining to Distribution of Liquefied Petroleum Gas and Superior Kerosene Oil in Bhutan, 1–10, online @ https://www.nationalcouncil.bt/assets/uploads/docs/download/2015/LPG_SKO_Study.pdf, 2014.

NSB. Statistical Yearbook of Bhutan. Thimphu: National Statistics Bureau, online @ http://www.nsb.gov.bt/publication/files/SYB_2018.pdf, 2018.

NSB. Statistical Yearbook of Bhutan. Public data, Thimphu: National Statistics Bureau, Royal Government of Bhutan, online @http://www.nsb.gov.bt/publication/files/yearbook2015.pdf, 2015.

NSB. Statistical Yearbook of Bhutan. Public data, Thimphu: National Statistics Bureau, Royal Government of Bhutan, online @ http://www.nsb.gov.bt/publication/files/yearbook2016.pdf, 2016.

NSB. Statistical Yearbook of Bhutan. Public data, Thimphu: National Statistics Bureau, Royal Government of Bhutan, online@http://www.nsb.gov.bt/publication/files/yearbook2017.pdf, 2017.

NSB. Statistical Yearbook of Bhutan. Public data, Thimphu: National Statistics Bureau, Royal Government of Bhutan, online @http://www.nsb.gov.bt/publication/files/yearbook2014.pdf, 2014.

NTGMP. National Transmission Grid Master Plan of Bhutan. Transmission Master plan, Thimphu: Department of Hydropower & Power Systems,online @ http://www.moea.gov.bt/wp-content/uploads/2018/11/National-Transmission-Grid-Master-Plan-2018.pdf, 2018.

RSTA. Road safety and trnasport authority. 2019. http://www.rsta.gov.bt/rstaweb/load.html?id=82&field_cons=MENU (accessed 2019).

Sadineni, Suresh B., Srikanth Madala, and Robert F.Boehm. "Passive Building Energy Savings: A Review of Building Envelope Components." Elsevier, 1–3, online @https://www.academia.edu/7818050/Passive_building_energy_savings_A_review_of_building_envelope_components, 2011.

Simanaviciene, Zaneta, Andzej Volochovic, Rita Vilke, Oksana Palekiene, and Arturas Simanavicius. "Research Review of Energy Savings Changing People's Behavior: A Case of Foreign Country." Elsevier, 1–5, online @http:www.sciencedirect.com, 2014.

SNV. Biogas progress report. Confidential project report, Thimphu:SNV Bhutan biogas project developer in Bhutan, 2016.

Tamang, Bharat. Overview of Bhutan-India Cooperation In the Power Sector. online @ https://sari-energy.org/oldsite/PageFiles/What_We_Do/activities/Bhutan/Overview_ of_Bhutan-India_Cooperation_in_the_Power_Sector.pdf, 2007.

Tshering, Sonam, and Bharat Tamang. Hydropower - Key to sustainable, socio-economic development of Bhutan. Thimphu, online @ https://www.un.org/esa/sustdev/sdissues/ energy/op/hydro_tsheringbhutan.pdf, 2004.

UN. Measurement Units and Conversion Factors. Part of the book chapter, UN manual, online @ https://unstats.un.org/oslogroup/meetings/og-04/docs/oslo-group-meeting-04–escm- ch04-draft1.pdf.

UN. World Urbanization Prospects. UN public report, United Nations economic windows, online @ https://population.un.org/wup/Publications/Files/WUP2018-Report.pdf, 2018.

UNDP. Undp Environmental Performance Reporting.Excel based country data on environ- ment performance, United Nations Development Program, n.d..

Ura, Karma, Sabina Alkire, Tshoki Zangmo, and Karma Wangdi. *An Extensive Analysis of GNH Index*. Research report, Thimphu: Centre for Bhutan Studies online @ http:// www.grossnationalhappiness.com/wp-content/uploads/2012/10/An%20Extensive%20 Analysis%20of%20GNH%20Index.pdf, 2012.

USAID. *South Asia Regional Initiative for Energy Integration*. 2019. https://sari-energy.org/ program-activities/cross-border-electricity-trade/country-energy-data/.

Wangchuk, Sangay, Stephen Siebert, and Jill Belsky. "Fuelwood Use and Availability in Bhutan: Implications for National Policy and Local Forest Management." *Springer Science+Business Media New York*,1–9, online @ http://citeseerx.ist.psu.edu/viewdoc/ download?doi=10.1.1.720.2411&rep=rep1&type=pdf, 2013.

World Bank. https://www.indexmundi.com/facts/bhutan/indicator/EN.ATM.CO2E.PC.2014.

World Bank. "Concept Note Bhutan Climate Fund." *Climate fund concept note*. Thimphu, online @ http://documents.worldbank.org/curated/en/971531538568130794/pdf/ Bhutan-Climate-Fund-Concept-Note.pdf, 2018.

Yangka, Dorji. "Carbon Neutral Policy in Action: The Case of Bhutan." 1–10, online @ https://www.researchgate.net/publication/329261835_Carbon_neutral_policy_in_ action_the_case_of_Bhutan, 2018.

Zangpo, Thukten. "Bhutan's first wind power project in Rubessa to power 600 house- holds." https://thebhutanese.bt/bhutans-first-wind-power-project-in-rubessa-to-power- 600-households/, 2014.

5 Sustainable Energy and Environmental Outlook: Indian Perspective

Asha Pandey

CONTENTS

5.1 Introduction 97
5.2 Global Energy Scenario 98
5.3 Indian Energy Scenario: Past and Present Trend 100
5.4 Environmental Scenario of India 102
5.5 Shift Toward Renewable Energy Resources 105
5.6 Renewable Energy Developments in India 106
    5.6.1 Solar Energy 108
    5.6.2 Wind Energy 110
    5.6.3 Biomass Energy 112
5.7 Sustainability in Renewable Energies: Environmental, Economic and Social Dimensions 113
    5.7.1 Sustainability in Solar Energy 113
    5.7.2 Sustainability in Wind Energy 114
    5.7.3 Sustainability in Bioenergy 114
5.8 Major Initiatives Toward Green Energies: Case Study 115
    5.8.1 Policy Framework and Government Initiatives 115
    5.8.2 Case Studies 117
        5.8.2.1 2G Ethanol Plant, Uttarakhand 117
        5.8.2.2 Muppandal Wind Farm 117
        5.8.2.3 Kamuthi Solar Farm 118
        5.8.2.4 Decentralized Energy Production in Indian Villages 118
5.9 Conclusion 118
References 119

5.1 INTRODUCTION

Energy is indeed the ability to do work, in the present scenario most of the human requirements cannot be fulfilled without energy. It is rightly considered as the backbone of the economy. Its availability and affordability is a crucial component of a nation's growth and development. Rapid urbanization and industrialization have given a quality of life to people dwelling in the cities. This desperate chase for quality of life has paid its toll on the economy in terms of expenditure on energy imports and economic inflation on one hand and contamination of food chains, scarcity of fresh air, water, and greenhouse gases (GHGs) on the other hand. India is often ranked as one of the fastest-growing economies of the world and is also often blamed for high GHG emissions and global warming. Fast economic growth is always accompanied by high energy demand; thus, India requires thinking about sustainable energy production. Keeping a dream of 100 percent rural electrification; India still has around 27 million households without any access to electricity as rural electrification deals with the electrification of only 10 percent household, public places, schools and health centers (SAUBHAGYA 2018) and around 780 million people still rely on biomass for cooking purposes (IEA 2017). Being a developing nation, India feels the pinch as it always remains on the pivot of balancing economic development and watching its carbon footprint. The major fraction of the country's finance is spent on fulfilling the energy requirements of the citizens, mainly crude oil imports. For instance, India spends approximately Rs. 8,81,282 crore on import of crude oil in the financial year which ended in March 2019. The demand for imports has increased by 42 percent if compared with the previous fiscal year. India's energy demand has outmatched the global demand growth in 2018 according to the International Energy Agency (IEA 2018). As per the present scenario, China, the United States, and India collectively account for nearly 70 percent of the rise in energy demand.

The upsurge in energy demand has led to a perplexing energy deficit in India and several other developing countries. Diminishing non-renewable resources and commitment toward reducing GHGs have raised the question of energy sustainability. The sustainability of energy resources is determined by its accessibility, affordability, and consistency. It has become the prima facie requirement of the energy market. Energy availability and security, monetary benefits and greenhouse gas reduction are the primary stimuli for the growth of the renewable energy market (Abolhosseini and Heshmati. 2014) in India. A gradual change in the energy mix has been observed in India with hydro, solar and wind energy taking its stand in the energy market. Government subsidies have encouraged and motivated people to switch to renewable energy. India has taken many new initiatives for energy sustainability. The present contribution focuses on the energy outlook of India and its leap toward renewable energy, the environmental scenario of the country, the potential and installed capacities for major renewable energies, policy framework, and the sustainability of energy resources in terms of social, economic and environmental spheres with relevant case studies.

5.2   GLOBAL ENERGY SCENARIO

Global energy production has been dynamic; the changes are observed not only in usage quantity, but also in the energy harnessing source. Until the 18th century, all countries primarily counted on traditional biomass for their energy needs. Renewable resources (wood and other organic matter) were used for cooking, heating, and other domestic purposes. However, a slight use of coal was noticed in the United Kingdom during the 19th century. Other resources such as oil, natural gas, and hydropower made their place as an energy resource in the early 20th century. The energy scenario

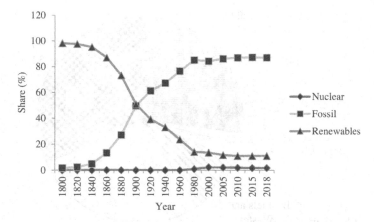

FIGURE 5.1   Changes in global energy mix during 1800 and 2018 (Source: BP Statistical Review of World Energy 2018).

was further diversified in mid-20th century; however, coal and oil always dominated the energy mix. Nuclear and other renewable energy resources gradually entered the energy framework which is still dominated by coal, oil, and natural gas. In past years, enormous changes had been noticed in the energy scenario of developed as well as developing nations. Figure 5.1 illustrates the scenario of changing the energy mix during 1800 and 2018.

The scenario changed with economic development; the developed nations with a stabilized economy, and technological advancement are now trying to move to sustainable, new, and renewable energy resources. However, developing nations aspiring for fast economic growth are trying to accomplish high energy demand by exploiting more conventional resources on one hand, and also trying to switch to renewable energy against global warming commitments on the other hand.

The global energy mix can be broadly classified into conventional, renewable, and nuclear energy. Coal, oil, and natural gas play a dominant place in conventional energy; however, hydropower holds a top place among renewable energies followed by solar, wind, and biomass. Nuclear energy, although harnessed, is not preferred due to its lethal implications. In search of clean fuel among the non-renewable resources, most of the developed nations have switched toward oil and natural gas rather than coal. IEA in 2018 has estimated a major change in energy mix due to volatility in oil prices and new energy policies. With an annual fluctuation of approximately 1.5–4.5 percent in global energy consumption, it is predicted that from the year 2018 to 2040 the energy demand will grow by 25 percent; the renewable energy market is expected to take a share of 40 percent in the energy mix in comparison to today's 25 percent share. Annual energy consumption (KWh) is recorded to be very high (>12,000 KWh/person/year) for the USA, Canada, Norway, UAE, Kuwait in comparison to India (1,181 KWh/person/year) (CIA World Factbook 2019). More than two-thirds of global electricity consumption is by top-ten consuming countries. Figure 5.2 illustrates the global energy mix based on fuel. Conventional energy resources still dominate the energy mix of the world.

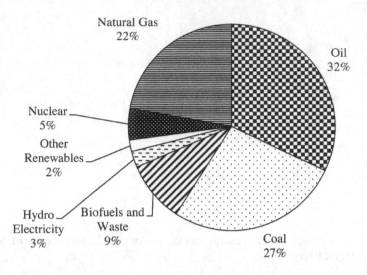

FIGURE 5.2   Global energy mix based on fuel (Source: IEA 2019).

In 2018, global primary energy consumption grew rapidly (2.9 percent) with natural gas and renewables taking the lead against previous years. Nevertheless, carbon emissions rose at their highest rate (2 percent) for 7 years. Total electricity generation rose by an above-average 3.7 percent, hovered by China (which accounted for more than half of the growth), India, and the US. The share of renewables in power generation increased from 8.4 to 9.3 percent (BP Statistical Review of World Energy 2019).

5.3   INDIAN ENERGY SCENARIO: PAST AND PRESENT TREND

Coal dominates the energy mix of India due to its availability in the country. Wood was conventionally used in all households but is becoming less accepted due to the availability of liquefied petroleum gas (LPG) for cooking purposes. India has witnessed a drastic change in energy mix during the last 20 years. Kerosene oil, fuel wood, crop residue, and cow dung cakes were essential components of the energy mix in the past but are absent in the present scenario.

The total primary energy supply (TPES) was 881.9 million tonne of oil equivalent (Mtoe) in 2017–2018 in which major contribution was made by fossil fuels. The details of total energy production as per present installed capacity and sector-wise electricity consumption are illustrated in Figure 5.3a and 5.3b, respectively. India's total final consumption (TFC) increased by 50 percent in the last decade from 2007 to 2018. Half the growth in this energy consumption came from the industrial sector, which accounted for 42 percent of TFC. The residential and agriculture sectors have high electricity consumption; however, fuel consumption is the third highest for the transport sector. Despite this, the per capita TPES and TFC of the nation are only 0.66 toe and 0.44 toe compared to the International Energy Agency (IEA) global average of 4.1 toe and 2.9 toe, respectively (Niti Aayog 2020).

India always faced an energy deficit due to demand being higher than generation. India has strived hard to reduce the energy deficit from 12 to 0.8 percent in the

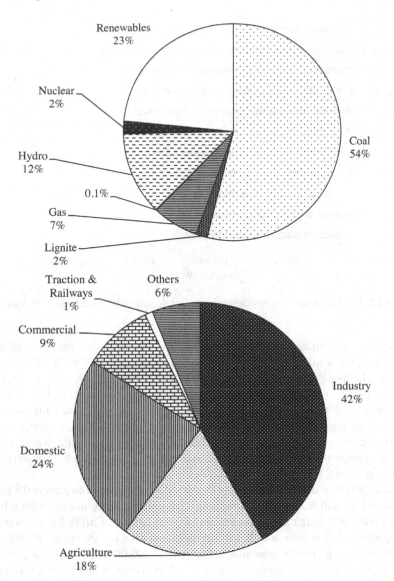

FIGURE 5.3 (a) Total energy generation (installed capacity) by fuel as of April 2020 (Source: MoP 2020). (b) Sector-wise electricity consumption in India (Source: CEA 2019).

last decade (2010–2020). Figure 5.4. illustrates the peak demand and supply (peak met) scenario of India in the past decade. Presently, India is one of the largest coal consumers in the world and also imports costly fossil fuel (Kumar and Majid 2020), which needs a rapid transition to renewable energy technologies to achieve sustainable growth and avoid catastrophic climate change.

The energy mix of India is gradually changing by exploring resources in different parts of the country. India is bestowed with diverse geographical and climatic

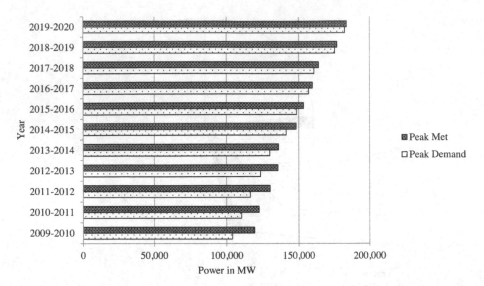

FIGURE 5.4 Reduction in energy deficit in the last decade (2010–2020) (Source: MoP 2020).

features that add uniqueness not just to its appearance, but also to energy availability. Apart from the availability of coal in many parts of India (Bihar, Jharkhand, West Bengal, Madhya Pradesh), the southern coastline (7561 km) provides the potential for wind energy, wave, tidal and ocean thermal energy conversion (OTEC). Western and central India give a good perspective for solar energy harnessing, while the mountainous regions of Himachal Pradesh and Uttarakhand bestow immense scope for hydro-electric power. Puga Valley in Ladakh, Himalayas, and the west coast also provide opportunities to harness geothermal energy. Nature has also conferred high biomass potential to all parts of India.

India makes a share of 17 percent of the world's population with merely 0.8 percent reserves of oil and natural gas resources. To maintain energy needs, India relies on costly oil imports from different countries (Energy Statistics 2019). Based on the past energy demand, it is expected that in the next two decades, by the year 2030, India should increase its power-generating capacity by 4,00,000 MW (presently 3,01,965 MW) to provide surplus energy. As per the projected economic growth levels, energy demand is expected to continuously rise, which in turn has drawn attention to the importance of energy security. The availability of energy, its access to people, and its affordability ensure *energy security*. India is now focusing on a sprint toward renewable energy for curbing energy security and sustainability. One more facet of this resolution is India's vulnerability to climate change and its global commitments.

5.4   ENVIRONMENTAL SCENARIO OF INDIA

Coal, oil, and natural gas are contributing one-third of global greenhouse gas emissions. It is now essential to economically grow with sustainable energy resources. Climate change might also change the ecological balance in the world. Global

Climate Risk Index 2020 report was published after COP 25 ranked India as the fifth most climate-vulnerable country out of 181 countries studies for quantifying the impacts of climate change through economic losses (The Economic Times 2019). Japan, Philippines, Germany, and Madagascar are preceding India with even higher vulnerability. India reported 2,081 deaths in 2018 due to extreme weather events caused by climate change-cyclones, heavy rainfall, floods, and landslides. Although, in 2019 India recorded just 9 disasters of 93 that occurred in Asia but accounted for 48 percent of deaths (State of India's Environment 2020).

Power production is the leading cause of CO_2 emissions in India, followed by industries and transport. CO_2 emission of the country based on sectors and fuel usage is illustrated in Figure 5.5a and 5.5b.

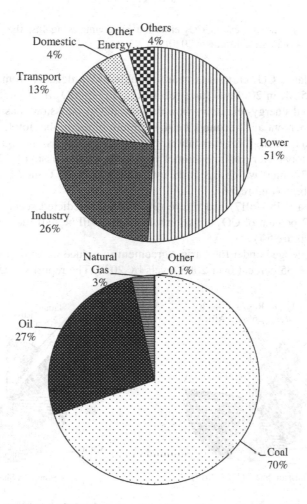

FIGURE 5.5 (a) India's CO_2 emission based on sectors (Source: Niti Aayog 2020). (b) India's CO_2 emission based on fuel usage (Source: Niti Aayog 2020).

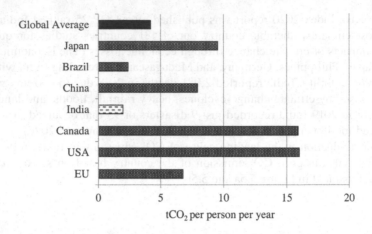

FIGURE 5.6 Annual per capita CO_2 emission by countries against the global average (Source: State of India's Environment 2020).

Energy-related CO_2 emissions nearly doubled in a decade, from 1,022 Mt in 2004 to 2,015 Mt in 2014 and have reached 2,162 Mt CO_2 in 2017. The country's share of global energy-related carbon dioxide (CO_2) emissions has increased by more than 2 percent and accounts for 4.4 percent of the global total. Although the CO_2 emissions are high for the nation, per capita emissions are significantly low against the global average. The annual per capita CO_2 emission by India in 2018 was only 1.97 tonnes which is significantly lower than the United States, Canada, Japan, and China (Figure 5.6).

As per past (1751–2017) and future (2019–2030) predicted scenario, India will contribute 8.2 percent of CO_2 emission which still will be half the fair share of its population (Figure 5.7).

India has pledged under the Paris Agreement to reduce the emission intensity of its GDP by 33–35 percent over 2005 levels by 2030. The report stated that India is

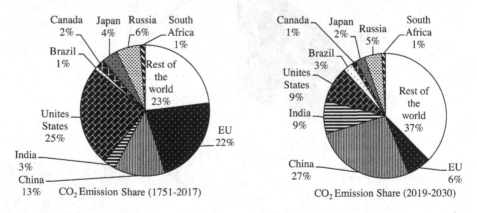

FIGURE 5.7 Comparative past (1751–2017) and future (2019–2030) CO_2 emission scenario (Source: State of India's Environment 2020).

on track to exceed this target by 15 percent. Besides, India has committed ensuring 40 percent of its installed power capacity from non-fossil sources by 2030 with a target of 175 GW renewable energy productions. This also helps India in beating another challenge of energy-related air pollution that is becoming catastrophic in India. The capital state of India, Delhi, has witnessed several episodes of smog (public health emergency) and is globally termed as a *hot gas chamber.*

India is proactively on the path of achieving the United Nations Sustainable Development Goals (SDGs) Agenda 2030. The United Nations Development Programme adopted 17 SDGs as a universal call to end poverty, protect the planet, and ensure that all people enjoy peace and prosperity by 2030. The goals can address both environment-related problems of climate change, SDG13, and air pollution, SDG3. The government of India (GoI) has adopted comprehensive and stringent rules for the power and transport sectors to tackle the problem of air pollution. SDG13 is being strictly adhered to by various policy agendas, reforms, and schemes. India is making good progress in its key targets for SDG7, namely to ensure access to affordable, reliable, sustainable, and modern energy for all. The target includes access to electricity and clean cooking. Some major initiatives successfully implemented are summarized in the later sections.

5.5   SHIFT TOWARD RENEWABLE ENERGY RESOURCES

Renewable energy resources are the resources that replenish in a short period and thus are considered renewable unlike conventional resources (coal and petroleum). The resources include energy from the sun, wind, hydro (water), waves, tides, ocean currents, and biomass, etc. All resources mentioned are free from pollution and can help in curbing the problem of global warming. The sources such as solar and wind energy have already made their place in energy market across the globe. Depending on their availability across various geographical locations, they can be harnessed and utilized for energy production.

Renewable energy has always been an integral part of Indian households. Traditional biomass for cooking, hydro-based grain grinding in Uttarakhand (locally called *Gharat,* Figure 5.8), wind-based irrigation system in Punjab, etc. were used long ago, but none of it was utilized in electricity generation. Renewable energy though was also existing in the energy market but has been accelerated due to the fact that it has been understood that conventional energy resources are about to exhaust. In India, the primary stimuli for growth of renewable energy market are biomass availability, energy sustainability, and monetary benefits from GHG capture (Abolhosseini and Heshmati. 2014).

Historical promotion of renewable energy in India started in 1981 with setting up of a Commission for Additional Sources of Energy (CASE) under the Department of Science and Technology. It was followed by the formation of the Ministry of Non-conventional Energy Sources (MNES) in 1992 which was renamed to MNRE in October 2006 (Bhattacharya and Jana 2009). India is the only country to have a full-fledged ministry for renewable energy production and with the largest production of energy from renewable sources. As of 2019, approximately 35 percent of India's installed electricity generation capacity is from renewable sources.

FIGURE 5.8 Traditional grain-grinding machine, *Gharat*.

In the initial years, the renewable energy market focused on hydroelectric projects which shared a major segment in the energy mix of India. However, with the strong voluntary commitments toward the Kyoto Protocol in 1997, India leaped into solar and wind energy projects. With the release of the National Action Plan on Climate Change (NAPCC) in 2008, India highlighted 8 missions for mitigating climate change: National Solar Mission (NSM), National Mission for Enhanced Energy Efficiency, Green India Mission are a few to name emphasized on renewable energy production in India. India's allegiance toward Paris Agreement was candid as nation aims of achieving 40 percent of its total electricity generation from non-fossil fuel sources by 2030. The country is aiming for a more ambitious target of 57 percent of the total electricity capacity from renewable sources by 2027 as per the Central Electricity Authority's strategy blueprint (The Guardian 2016). Presently, hydro-electric power, wind energy, and solar energy are the key players in the Indian renewable energy market followed by biomass, waste to energy, and cogeneration.

India has globally announced its ambitious renewable energy targets and is determined to produce 175 GW of renewable energy by 2022, almost 430 GW by 2030, and 800 GW by 2040. The country has committed 40 per cent energy from renewable in its energy mix by the year 2040.

5.6 RENEWABLE ENERGY DEVELOPMENTS IN INDIA

The commitment and targets for renewable energy have been discussed in the prior sections of the chapter. The total potential for renewable power generation in India is estimated (as of March 2018) to be 10,96,081 MW. Renewable energy includes solar power potential of 7,48,990 MW (68.33 percent), the wind power potential

TABLE 5.1

Total Potential and the Installed Capacity of Renewable Energy in India

Renewable Energy	Total Potential (MW)	Total Installed Capacity (MW)
Wind Power	302251	37669.00
Solar Power – Ground Mounted	748990*	27930.32
Solar Power – Roof Top		2141.03
Small Hydro Power	19749	4593.15
Biomass Power (Biomass and Gasification and Bagasse Cogeneration)	20536	9778.31
Waste to Energy	2554	138.30

* total solar potential (ground-mounted and rooftop)
Source: MNRE 2019.

of 3,02,251 MW (27.58 percent) at 100 m hub height, small-hydro power (SHP) potential of 1,97,49 MW (1.80 percent), biomass power of 17,536 MW (1.60 percent), 5000 MW (0.46 percent) from bagasse-based cogeneration in sugar mills and 2,554 MW (0.23 percent) from waste to energy (MoSP 2019). To increase investment in renewable electricity in a cost-effective way, India has introduced national competitive auctions for wind and solar photovoltaic (PV). Total potential and installed capacity for grid-connected renewable energy of India are tabulated below (Table 5.1).

For India, 2017 was a flourishing year in renewable energy. The country witnessed an epic increase in renewable power capacity, steep fall in its cost, and whopping investment and advances in renewable technologies. Renewable energy progress in the transport sector remains slow yet bio-fuels provide most of the current renewable energy contributions, similarly solar and wind have also played a crucial role in installed grids in India (Global Status Report 2018).

Hydro-electric projects (HEPs) were billed to be the most sustainable, renewable alternative to coal-and gas-based electricity. However, since 2015, the Indian Government has stopped categorizing HEPs larger than 25 MW as renewable energy projects because of the negative social and environmental implications of these projects. Presently, large HEPs share 13 percent of energy in the Indian energy mix but were not considered in renewable energy. Conversely, in March 2019 Union Cabinet again included the projects under renewable energy to accomplish the pledge of zero carbon emissions.

The present renewable energy mix (excluding large HEPs) makes a total renewable share of approximately 21 percent in the overall energy mix (Figure 5.9). India targets to commission 175 GW of renewable energy capacity by 2022. As of March 2019, the total grid-connected installed capacity from renewable energy sources is recorded to be 77.64 GW (MNRE 2019).

The major renewable energy sources, their geographical locations, present generation scenario and future targets are discussed below.

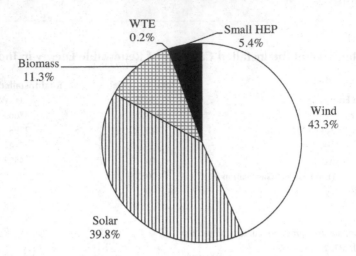

FIGURE 5.9 Renewable energy mix (installed capacity) of India (as of March 2020) (Source: MNRE 2019).

5.6.1 SOLAR ENERGY

The energy from the sun is defined as solar energy. Apart from providing many other sources of energy for us, the sun itself has an enormous amount of energy which is generated from nuclear fusion reaction taking place in it. The sun can annually produce approximately 23,000 TW of energy; if fully captured, it comes out to be much higher than the world's annual energy requirement. However, it is practically impossible to do that; therefore, we look for technologies that can trap the maximum amount of sun's energy.

Solar energy's harnessing potential mainly depends on geographical location, the sun's angle, latitude, and air mass number (thickness of atmosphere). The appropriate geographical location of India bestows about 300 clear and sunny days to it in a year. With a huge landmass, approximately 5000 trillion kilowatt-hours (kWh) of solar energy annually incidence on land makes it economical to harness this clean renewable source of energy. Based on the availability of land and solar radiation, the potential solar power in the country has been assessed to be around 750 GWp. Rajasthan, Karnataka, Gujarat, Maharashtra, and Tamil Nadu are a few states leading in solar power generation (Figure 5.10).

Solar energy has witnessed maximum growth in the past decade and has captured a major energy market. The increase in solar energy generation is illustrated in Figure 5.11.

India generates its solar energy under the National Solar Mission (NSM) of NAPCC, implemented on 11th January, 2010. The initial target for solar power development and deployment was 20 GW by 2022. The target was achieved before the due date, thus the Cabinet in 2015 revised its target to 100 GW. Solar energy in India is captured based on the availability in different parts of the country. The nation is working on both grid-connected and off-grid projects. The primary focus is on solar photovoltaic via the construction of solar parks, which are a large area of land developed with all necessary infrastructures and clearances for setting up solar projects. The capacity of the solar parks is generally 500 MW and above. Approximately 4 to

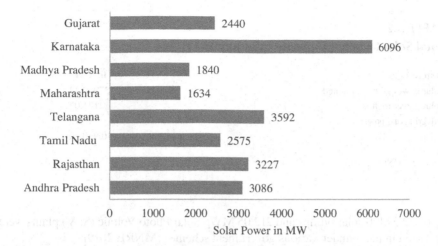

FIGURE 5.10 Top-most Indian states in grid-connected solar energy generation (Source: MNRE 2019).

5 acres of land per MW is required for setting up a solar park. The target of the solar park scheme is to develop at least 50 solar parks with an aggregate installed capacity of 40,000 MW of solar power by 2021–2022.

Apart from the solar park, solar rooftop photovoltaics and off-grid solar power are two main solar energy generation methods. Off-grid solar power deals with the deployment of solar street lights, solar pumps, solar power parks, and other solar applications to meet the electricity and lighting needs of the local communities/institutions/individuals in the rural areas. Under off-grid solar energy approximately 58,23,710 solar lanterns and study lamps; 17,15,214 solar home lights; 6,59,218 street

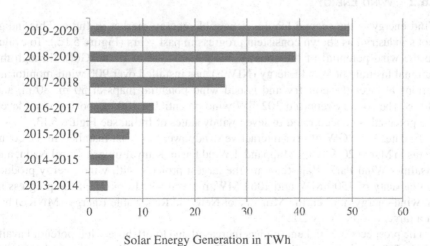

FIGURE 5.11 Increase in solar energy generation in past years (2013–2020) (Source: MNRE 2019).

TABLE 5.2
Total Solar Installed Capacity as of July 2019

System Type	In MW
Solar power ground mounted	27930.32
Solar power rooftop	2141.03
Off-grid solar power	919.15
Total	30,990.50

Source: MNRE 2019.

lights; 2,37,120 solar pumps and 212.05 MWp Solar Photo Voltaic (SPV) plants were installed in India under various government schemes (MNRE 2019).

Solar heating by the installation of a solar water heater is also prominent in various parts of the country. The total installed capacity of 30.99 GW was reported on July 2019. The solar energy generation breakup is given in Table 5.2.

In the last decade, 30 percent annual growth has been recorded in the solar energy market. The solar capacity of India has increased by 370 percent in the last three years (MNRE 2019). The country's solar installed capacity reached 29.55 GW as of 30 June, 2019. Major solar farm installation in Karnataka, Andhra Pradesh, Tamil Nadu, Rajasthan, and Gujarat has been the major initiative of India toward renewable energy aspirations. The capital cost per MW for installing solar power plants is lowest globally. The solar tariff has reached a record lowest at Rs. 2.43 (US$0.037) per unit in December 2017 and 2018 (MNRE 2019).

5.6.2 WIND ENERGY

Wind energy is the second leading renewable energy market in India. This indigenous industry has shown consistent progress in past years (Figure 5.12.). To evaluate the wind potential of the country, the government of India (GoI), through the National Institute of Wind Energy (NIWE), has installed over 900 wind-monitoring stations all over the country and issued wind potential maps at 50 m, 80 m, and 100 m. The country reported 302 GW wind potential at 100 m above ground level. The potential is concentrated in seven windy states of India, see Figure 5.13.

To date, 35.62 GW of grid-interactive windpower has been installed in the country (as in March 2019) with Muppandal Wind Farm, Kanyakumari, Tamil Nadu; and Jaisalmer Wind Park, Rajasthan are the largest projects with wind-energy production capacity of 1500 MW and 1064 MW, respectively. Looking at the progress in the wind-energy market, the Ministry of New and Renewable Energy (MNRE) has set a target of 60 GW by 2022.

The prospects of wind energy has increased due to offshore wind potential available in 7600km long coastline of India. Government of India has notified *National Offshore Wind Energy Policy'* on 6th October, 2015 to harness the untapped wind potential in offshore locations. The first offshore wind-energy project of 1.0 GW

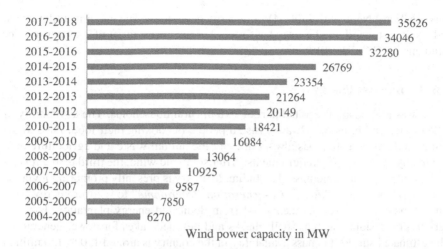

FIGURE 5.12 Increase in installed windpower capacity of India (2005–2018) (Source: MNRE 2019).

capacity is planned in Gujarat. India is estimated to have an offshore wind-energy potential of around 70 GW in parts along the coast of Gujarat and Tamil Nadu.

India is bestowed with high wind potential due to its unique topography and climate. In the last few years, wind-energy production has surpassed all other renewable energies in India. India has recorded the total installed wind-power capacity of 35.28 GW which makes it the fourth largest installed wind-power capacity in the world. Tamil Nadu, Maharashtra, and Gujarat states have exhibited the highest potential for wind power. The largest wind farm in India is the Muppandal Wind

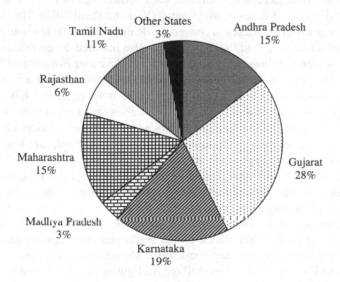

FIGURE 5.13 Ranking of Indian States in wind-energy potential (Source: MNRE 2019).

Farm in Tamil Nadu with 1500 MW of energy generation. The major installations in India's drive toward wind energy have recorded the lowest tariff of 2.44 per unit for wind energy (MNRE 2019).

5.6.3 BIOMASS ENERGY

Wood was always an integrated part of Indian rural households. The availability of fallen twigs and branches always assisted people in cooking their food. Also, India being an agrarian nation has always utilized the animal waste and agriculture residue in day-to-day energy requirements. After solar and wind, the third prime renewable source in India is biomass. The Indian Ministry is presently promoting it via the *"Biomass Power and Bagasse Co-generation Programme"* to recover energy from various biomass feedstocks, i.e., wood from dedicated energy plantation, bagasse, agricultural residues such as shells, husks and de-oiled cakes for power generation. The estimated surplus biomass availability in the country is around 120 to 150 million tonnes per annum which can give sustainable biomass for energy generation. The biomass energy potential is estimated to be 26000 MW which includes 18000 MW from agricultural and agro-industrial residues and an additional potential of 8000 MW from bagasse cogeneration in sugar mills. Over 500 biomass power and cogeneration projects with an aggregate capacity of 9103.50 MW have been installed in the country as of March 2019. These projects have been commissioned mainly in the states of Tamil Nadu, Uttar Pradesh, Karnataka, Andhra Pradesh, Maharashtra, Chhattisgarh, West Bengal, and Punjab.

The cultural ecology of India encourages cattle breeding in every individual rural family. Biogas or Gobar (cow dung) gas has been practiced earlier in many villages of India. MNRE technically stimulated the same through *"New National Biogas and Organic Manure Programme"* (NNBOMP) in 2018 based on cattle dung and biodegradable waste in remote, rural and semi-urban areas of India. The scheme is implemented across 13 states of the country (Rao and Baral 2013). The scheme can support remote off-grid villages to convert cattle dung and organic wastes into clean gaseous fuel for cooking and lighting. MNRE also initiated decentralized *"Biogas-based Power Generation and Thermal Energy Applications Programme"* (BPGTP) in 2017 for the deployment of medium-sized biogas plants in the rural, semi-urban areas of the country, having captive power generation in the range of 3 KW to 250 KW to meet captive power needs of farmers, small diaries and rural industries. The plant can produce 30 m^3 to 2500 m^3 per day of biogas using organic biodegradable wastes from various sources such as cattle dung/animal wastes, food, and kitchen waste, poultry dropping waste, agro-industry waste, etc. As of March 2019, 389 biogas-based projects with total biogas generation of 87,990 m^3 per day corresponding to the power generation capacity of 8.951 MW have been installed in the country.

MNRE is promoting a comprehensive *Biomass Gasification Programme* (BGP) for producing electricity using locally available biomass resources such as small wood chips, rice husk, *arhar* stalks, cotton stalks and other agro-residues in rural areas. The program aims to meet the captive electrical and thermal needs of the unmet demand for electricity of the village for lighting, water pumping, and micro-enterprises. Unnat Chulha Abhiyan (UCA) scheme has been also implemented to promote biomass cookstoves. The country has deployed around 51641 Improved Biomass Cookstoves under the UCA Program, which includes 50,322 family type and 1319 community type. To see ensured positive response, renewable purchase obligation (RPO) has been mandated to all State Electricity Regulatory Commissions to promote renewable energy. The Ministry of Power notified the uniform RPO trajectory that seeks 21 percent RPO (10.5 percent non-solar and 10.5 percent solar) by 2021–2022.

Waste to energy (WTE) and biofuels such as bioethanol and biodiesel are an additional dimension of exploring biomass energy in India. WTE not being very fruitful has provided 147.64 MW of installed capacity as of March 2020. Fuel blending, i.e., bioethanol in petrol and jatropha seed oil in diesel, has been implemented under Ethanol Blended Petrol Program, National Biodiesel Mission, and Biodiesel Blending Programs. An ambitious goal of 30 percent blending is set by 2020. However, 7.2 percent ethanol blending with petrol was achieved in 2018–2019; biodiesel blending in diesel remained only 0.14 percent. The blending programs can help India in reducing its dependence on oil imports (Singh 2019).

5.7 SUSTAINABILITY IN RENEWABLE ENERGIES: ENVIRONMENTAL, ECONOMIC AND SOCIAL DIMENSIONS

Sustainable development is imperative globally and thus, the UN General Assembly has declared the years 2014–2024 to be the *"Decade of Sustainable Energy for All"*. The word sustainability has become the jargon of the present generation. Sustainability is generally described as the availability of resources for present and future generations. Energy sustainability is also bracketed similarly. But the availability, accessibility, and affordability need to be clearly understood when talking about energy sustainability. Sustainability should also keep a watch on the long-term impacts of present innovative resources. The development and installation of new and renewable energy resources should not take its toll on the future generation. The social angle of sustainability, which is often neglected, plays the most important role in energy sustainability. This section describes the sustainability issues associated with the three major renewable energy resources of India, which are presently neglected but can create havoc in the future.

The decentralized approach is constantly neglected in energy scenarios which is the most important pillar of sustainability. The utilization of resources within one's geographical location always maintains the appropriate ecological balance. The decentralized system also reduces the transportation loss of energy.

5.7.1 SUSTAINABILITY IN SOLAR ENERGY

Solar power is indeed a clean source of energy which is having enormous potential in many parts of the world. Although it is renewable energy, it is largely made up of non-renewable resources. Its limited lifespan and getting obsolete shortly raise a question on its sustainability. The manufacturing of solar panels involves the usage of many substances which are not eco-friendly. For instance, nitrogen trifluoride is a byproduct of electronic manufacturing often used in solar cells; it is a green-house gas with 17,000 times higher global warming potential than carbon dioxide. Advanced solar cells are using cadmium in manufacturing which is a toxic heavy metal (Dahren 2016).

Solar panels offer a risk of lead, cadmium, and other toxic chemicals contamination in the environment. Approximately 90 percent of photovoltaic (PV) modules are made up of glass, which often cannot be recycled as float glass due to impurities. The glass used is polyvinyl chloride (PVC) consists of many impurities such as plastics, lead, cadmium, and antimony. Many environmentalists articulate that solar panel disposal in regular landfills is not recommended due to the risk of breaking of modules and toxic materials leaching into the soil. One of the local concerns and threats being posed by solar panels is the washing of cadmium with the rainwater. The stand-alone system needs big batteries whose storage and disposal can be a real threat soon. Sustainability also looks for sustainability in the future which looks vague for solar energy (Kazmeyer 2017).

5.7.2 SUSTAINABILITY IN WIND ENERGY

Wind energy is certainly another clean and renewable energy but does offer some unreliability elements. The foremost important issue is the unpredictable nature of winds which force the usage of auxiliary fuel. The construction of the tower, transport, and installation of huge blades and maintenance at such a great height often causes high monetary constraints (Natarajan 2014); thus hampeing the aim of sustainability. Wind turbines harm the avian population. Studies conducted in wind farms of Gujarat and Karnataka, India have reported a total of 47 bird carcasses (11 species) in 3 years in Kutch, Gujarat and 7 carcasses (3 species) in 1 year were recorded at Davengere, Karnataka (Kumar et al. 2019). Noise pollution and visual interferencc are also the inevitable impacts of wind energy which create stress and discomfort to people residing in those areas (Rahman et al. 2011). On the contrary, earlier researchers believe in dual sustainability of wind energy, which means the achievement of environmental and financial sustainability simultaneously (Welch and Venkateswaran 2009). With strong sustainability goals and targets, wind energy has been partially criticized in economic, social, and environmental spheres of sustainability.

5.7.3 SUSTAINABILITY IN BIOENERGY

Looking for all the dimensions of sustainability, bioenergy offers positive outcomes in the economic, environmental, and social spheres of sustainable development. The direct involvement of the rural population and close access to biomass offer social sustainability to the resource. However, many researchers working on agriculture have contrary views on the sustainability of bioenergy due to food security. The major issue on the sustainability of bioenergy is raised on land availability. Land being a non-renewable resource needs to be used judiciously and carefully for agriculture, energy, or any other utilization. The use of wastelands and marginal land for bioenergy production solves the issues of land sustainability (Pulighe et al. 2019).

The management of bioenergy in the supply chain is a complex process and is opening its way forward. Biomass utilization is an economic alternate energy that offers benefits to the social sphere by providing monetary and livelihood opportunities, and to the environmental sphere by reducing GHGs much more than any renewable energy. The emission of pollutants by burning biomass has been resolved by the innovation of gasifiers and many other thermo- and biochemical technologies.

5.8 MAJOR INITIATIVES TOWARD GREEN ENERGIES: CASE STUDY

5.8.1 Policy Framework and Government Initiatives

India has proactively focused on improving energy scenarios to obtain fast and sustainable growth in the country. The GoI has ventured into a pronged approach to cater to the energy demand of its citizens while ensuring a minimum rise in CO_2 emissions to protect the environment. The initiatives are equally commenced on generation-distribution side, and energy efficiency. The promotion of renewable energy mix with stringent renewable energy targets, together with improvement in the efficiency of thermal power plants, is implemented successfully in the country. Innovative energy policies and reforms under the Energy Conservation Act of 2001 are also jointly in action. Clean cooking initiatives are also extensively implemented to address the problem of air pollution.

To provide continuous power supply to rural India, in 2015, *Deendayal Upadhyaya Gram Jyoti Yojana* (DDUGJY) with a total budget of USD $10.8 billion was introduced followed by the *Saubhagya Scheme* commonly known as *Pradhan Mantri Sahaj Bijli Har Ghar Yojana* with a total budget of USD $1.8 billion was implemented for last-mile connectivity to households. The energy needs of urban areas of India are also facilitated with the Integrated Power Development Scheme (IPDS). The schemes provide budgetary support (grants) to state Government Distribution Companies (DISCOMs) (Niti Aayog 2020). Subsidy in electricity is provided not only to domestic consumers, but also to the agriculture sector consumers who are below the poverty line (BPL) under the National Electricity Policy of the country. To ensure that the benefits reach citizens, the government subsidy is directly provided through Direct Benefit Transfer via the Aadhaar identity system.

Initiatives are not only taken to provide electricity, but also to promote clean cooking, i.e., LPG to reduce the exposure to indoor air pollution from burning wood or dung, and also the time needed for collecting fuel wood or unsustainable biomass. Under Pradhan Mantri Ujjwala Yojana (PMUY) in 2016, 50 million LPG connections were distributed to women of BPL families. A heavy subsidy was also introduced which sharply increased to 80 million connections in 2019. LPG cylinders for domestic utilization are sold at market price; however, the subsidy for the same is directly credited to a consumer's accounts under *Pratyaksh Hanstantrit Labh* (PAHAL) scheme. Key to the scheme's success has been the Aadhaar identity system, which links subsidy payments to bank accounts, and the better targeting of subsidies directly to women, which has increased women's financial inclusion and access to clean cooking. Presently, 9.75 crore LPG consumers are facilitated across the country. The GoI also launched a voluntary program named *"#GiveItUp"* campaign which aims to motivate LPG users to surrender their subsidy if they can afford to pay the market price for LPG (Niti Aayog 2020).

Under energy efficiency, *Unnat Jyoti by Affordable LEDs for All* (UJALA) scheme was launched in 2015 which aims to promote efficient use of energy at the residential level. Under this scheme, 20 W LED tube lights with 50 percent more energy efficiency and ceiling fans with 30 percent energy efficiency were made available to consumers at half price. Awareness of consumers about energy-efficient home appliances, with high initial cost but long-term benefits, was also accomplished under the program.

As 60 percent of the Indian population still depends on agriculture, energy efficiency was also a prime requirement. *Pradhan Mantri Kisan Urja Suraksha evam Utthaan Mahabhiyaan* (PM KUSUM) scheme worked on the installation of solar pumps to reduce the energy demand of the agriculture sector.

The GoI, with a broad vision, has energy sustainability as its top priority. The government has embarked on an ambitious policy to boost renewable electricity, with a target of 175 GW capacity by 2022. Under its Nationally Determined Contribution (NDC), India targets a share of non-fossil-based capacity in the electricity mix of more than 40 percent by 2030. Recently, the GoI has indicated ambitious new targets for renewables capacity in the region of 450 GW by 2030. India's energy sector is transitioning to greater sustainability. As a result of the country's proactive and sustained actions on climate change mitigation, the emissions intensity of India's GDP has reduced by around 20 percent over the period 2005–2014.

The National Action Plan on Climate Change (NAPCC) – India has distinctively shown its commitments toward climate change and its adaptation and mitigation strategies by announcing the release of India's first comprehensive report outlining existing and future policies and programs addressing climate mitigation and adaptation. Every state of India made its contribution via its State Action Plan on Climate Change (SAPCC) which presented the ground level problems associated with climate change in a particular geographical area. The NAPCC identifies measures that promote development objectives while also yielding to co-benefits for addressing climate change effectively and pledges that India's per capita GHG emissions will at no point exceed those of developed countries even in pursuing development objectives. The NAPCC identifies eight core *National Missions* for the country:

* *National Solar Mission*
* *National Mission for Enhanced Energy Efficiency*
* *National Mission on Sustainable Habitat*
* *National Water Mission*
* *National Mission for Sustaining the Himalayan Ecosystem*
* *National Mission for a "Green India"*
* *National Mission for Sustainable Agriculture*
* *National Mission on Strategic Knowledge for Climate Change*

Based on India's stated policies and its renewable energy targets of 175 GW by 2022, almost 430 GW by 2030 and 800 GW by 2040, the International Energy Agency (IEA) has predicted that India can increase the share of modern renewable to 17.5

percent by 2040 and share of traditional biomass can be reduced from 22 to 8 percent by 2040. Electricity generation from renewables is also predicted to grow from 18 to 45 percent by 2040.

5.8.2 CASE STUDIES

India has ranked itself as one of the largest energy producers through renewable energy. As of September 2019, the installed renewable energy capacity is 81.3 GW, of which solar and wind comprise 30.70 GW and 36.75 GW, respectively. Biomass and small hydropower constitute 9.81 GW and 4.6 GW, respectively (Energy Statistics 2019). In the Paris Agreement, India has targeted to generate 40 percent of its energy from renewable resources by 2030, which now looks to be an achievable goal. Strong commitments toward global warming, together with depleting conventional energy resources and hefty monetary loss due to energy (oil) import, are the three prime motivators for India's sprint into the renewable energy market. Some of the major Indian initiatives are discussed in the following section.

5.8.2.1 2G Ethanol Plant, Uttarakhand

India inaugurated its first 2G ethanol plant in Kashipur, Uttarakhand on 22 April, 2016 (Earth Day). The technology was developed by the DBT-ICT Centre for Energy Biosciences in association with India Glycols Private Limited. The feed-stock used in ethanol production is hardwood chips, cotton stalk, soft bagasse, and rice straw. The technology is a continuous process and can convert biomass feed to alcohol in 24 hours. The facility is scalable from 100 tons of biomass/day to as much as 500 tons/day making it flexible with feedstock availability. The technology was initially tested on a pilot scale (1 ton per day) at the India Glycols site. The plant is targeted to have an annual capacity of 7,50,000 liters. The project cost was $5.28 million. The initiative is inevitable keeping in mind the mandate of 5 percent green biofuel into gasoline as per the National Biofuel Policy drafted in 2009. Presently ethanol blending is approximately 6.2 percent (was 4.2 percent in the previous year) is expected to increase sharply. Similar projects are expected to be replicated in Odisha by 2020. Researchers have estimated availability of varied feedstock across many states of India, namely, rice straw in Punjab and Haryana; cotton and castor stalk in Gujarat and Maharashtra; bagasse and sugar cane trash in UP, Punjab, Tamil Nadu, and Maharashtra; palm empty fruit bunches in Andhra Pradesh; and bamboo in Assam, Bengal, and Orissa along with municipal solid waste (MSW) which can provide the required feedstock (Biofuel Digest 2016).

5.8.2.2 Muppandal Wind Farm

Muppandal Wind Farm is the largest operating wind farm in India situated in the Kanyakumari district of Tamil Nadu. The farms come under Tamil Nadu Energy Development Agency with an installed capacity of 1500 MW, making it one of the world's largest onshore wind farms. A total of 483 windmills are installed on the farm. The state is known to have high wind energy potential due to the tunneling effect of SW monsoon. The average wind speed in 19 – 25 km/hr with a power density of 373. Unlike other wind farms, Muppandal Wind Farm gives high power density and has completely changed the energy scenario of the state (Natarajan 2014).

5.8.2.3   Kamuthi Solar Farm

Another green initiative taken by India was one of the largest solar power parks set up at Kamuthi Ramanathapuram, Tamil Nadu. It is a photovoltaic power station spread over an area of 2500 acres ($10km^2$). The plant has an installed capacity of 648 MW which can fulfill the energy requirement of approximately 7,50,000 people. The plant incurred a high expenditure of Rs. 4550 crore (USD $710 million in 2018) from Adani Green Energy Ltd. (Adani Group). The solar plant consists of 2.5 million solar modules and other related infrastructure (inverters, transformers, etc.). It is the world's largest solar power harnessed at a single location (Reid 2017).

5.8.2.4   Decentralized Energy Production in Indian Villages

India, apart from being a major green energy player in the world, is also focused on decentralized energy initiatives. Based on the feedstock availability, many initiatives have been taken in biogas production as well as biomass gasification for rural electrification. One of such small but significant initiative was taken by the villagers of the Hanumanthanagara (58 households) and Hosahalli (35 households; unelectrified) village situated in Tumkur District of Karnataka (100 km from Bangalore). The biomass gasifier-based decentralized power generation system was commissioned in 1988 in Hosahalli (Ravindranath et al. 1990, 2004) and an identical capacity system was commissioned in Hanumanthanagara in 1997 in collaboration with Centre for Application of Science and Technology for Rural Areas, Indian Institute of Science, Bangalore. A 20 kW biomass gasifier was initially installed in Hosahalli and was later replicated in Hanumanthanagara. The plant had a conversion efficiency of 21 percent from wood to electricity. The energy was utilized for lighting the houses, pumping drinking water, running flour mills, and irrigation purposes. Similar projects on rural electrification in villages (Madhya Pradesh, Gujarat, Uttarakhand, etc.) are carried over in many parts of India.

India has witnessed many such unknown projects on green energy (decentralized and large scale). The mandatory usage of refuse-derived fuel by industries, i.e., cement kilns and installation of many wastes to energy projects are some of the highlighting initiatives of the nation. Despite the enormous population and lower economic stability, the country is taking all possible initiatives to be green and sustainable.

5.9   CONCLUSION

India is a huge landmass with various geographical features. Many of the places are remote with no physical access. The collection of data on total final energy consumption becomes a major challenge for most of the State Ministries. The transition of solar and wind energy needs an excess of land, in such cases, land acquisition from the local population creates a lot of agitation and upheaval. Land acquisition is more problematic in on-shore wind farms. Most of the labor force of the country is not skilled and depends on local resources for its sustenance and survival. Lack of funds and investments often increase the capital cost, thus causing a delay in project development. Aspiration of a lean carbon economy also affects the gross domestic product (GDP) of the nation as the initial cost of all renewable projects is high. Technology development and feedstock logistics such as the collection of waste and residues are also one major challenge in biomass energy development.

Energy is indeed the pre-requisite for a comfortable life. The energy availability is important but the impacts should not ruin the environment and burn a hole in people's pockets. India is a developing nation with 22 percent of its population still living below the poverty line, i.e., Rs. 32/day in rural areas and Rs. 47/day in urban areas (Planning Commission, 2015). Energy availability, affordability, and uninterrupted access play an important role in sustainability. The energy resources which fulfill these targets should be available for a long time without any fear of depletion.

After the hue and cry of depleting conventional energy resources and the increasing GHGs and global temperature, every country is trying hard to shift its energy mix toward new and renewable energy resources. For instance, the shift toward electric vehicles is one of the major steps toward GHG and pollution prevention, but the availability of electricity could raise a question on sustainability if not solely from renewable resources. Also, the introduction of new technology should not hamper social comfort and well-being. Present momentum for solar, wind and hydro-resources are ignoring the future impacts on the environment and society. This *myopic vision* of energy generation and utilization will hinder sustainability shortly. New issues of sustainability will take their roots overriding the old ones. The unreliability and unequal distribution of energy resources can lead to environmental degradation and low quality of life.

The sustainability manifests in choosing resources based on a geographic and economic background. For example, the dependence on the waste to energy plants for electricity cannot be a viable option for a country such as India which produces 60 percent of wet waste (from the kitchen) with low calorific value. The transport of energy often increases the cost and maintenance in a country with long geographical boundaries. Remote rural India which needs electrification cannot wait for grid connection and thus needs decentralized energy production initiatives from the government. It is time to change India's vision on energy and discover not only new energy resources for us and future generations, but also such resources which do not cause health hazards and waste management issues for our future generation.

REFERENCES

Abolhosseini, S. and A. Heshmati. 2014. The main support mechanisms to finance renewable energy development. *Renewable and Sustainable Energy Reviews*. 40(issue C): 876–885.

Bhattacharya, S. C. and C. Jana. 2009. Renewable energy in India: Historical developments and prospects. *Energy* 34: 981–991.

Biofuel Digest. 2016. Indian cellulosic ethanol technology set for debut at demonstration scale in Uttarakhand. http://www.biofuelsdigest.com/bdigest/2016/04/21/indian-cellulosic-ethanol-technology-set-for-debut-at-demonstration-scale-in-uttarakhand/

BP Statistical Review of World Energy. 2018. (eds 67th). https://www.bp.com/content/dam/bp/business-sites/en/global/corporate/pdfs/energy-economics/statistical-review/bp-stats-review-2018-full-report.pdf

BP Statistical Review of World Energy. 2019. (eds 68th) https://www.bp.com/content/dam/bp/ business-sites/en/global/corporate/pdfs/energy-economics/statistical-review/bp-stats-review-2019-full-report.pdf.

Central Electricity Authority (CEA). 2019. Ministry of Power, Government of India, New Delhi.

CIA World Factbook, 2019. List of countries by electricity consumption. https://en.wikipedia.org/wiki/List_of_countries_by_electricity_consumption, accessed on 2 June 2020.

Dahren, B. 2016. Renewable energy – not always sustainable. https://phys.org/news/2016-10-re newable -energy-sustainable.html.

Energy Statistics. 2019. Ministry of Statistics and Programme Implementation, Government of India, New Delhi.

Global Status Report. 2018. Renewables 2018. Renewable energy policy network for the 21st century (REN 21). http://www.indiaenvironmentportal.org.in/files/file/Renewables20 201820 Global20Status20Report.pdf

International Energy Agency (IEA). 2017. World Energy Outlook Special Report (October). https://www.iea.org/reports/energy-access-outlook-2017.

International Energy Agency (IEA). 2018. World Energy Outlook (November). https://www.iea.org/reports/world-energy-outlook-2018.

International Energy Agency (IEA). 2019. International Energy Outlook (November). https://www.eia.gov/outlooks/ieo/pdf/ieo2019.pdf

Kazmeyer, M. 2017. Myths of Solar Power. Sciencing. https://sciencing.com/myths-solar-power-22753.html

Kumar, A., N. Kumar, P. Baredar and A. Shukla. 2015. A review on biomass energy resources, potential, conversion and policy in India. *Renewable and Sustainable Energy Reviews*. 45: 530–539.

Kumar, C. R. and M. A. Majid. 2020. Renewable energy for sustainable development in India: current status, future prospects, challenges, employment, and investment opportunities. *Energy, Sustainability and Society*. 10(2).

Kumar, S. R., V. Anoop, P. R. Arun, R. Jayapal and A. M. S. Ali. 2019. Avian mortalities from two wind farms at Kutch, Gujarat and Davangere, Karnataka, India. *Current Science* 116(9): 1587–1591.

Ministry of New and Renewable Energy (MNRE). 2019. Annual Report, 2018-2019. Government of India, New Delhi.

Ministry of Power (MoP). 2020. Power sector at a Glance All India. Government of India, New Delhi.

Ministry of Statistics and Programme Implementation (MoSP). 2019. Energy Statistics 2019 (26th Issue). Government of India, New Delhi.

Natarajan, V. K. 2014. Sustainability in India through wind: A case study of Muppandal wind farm in India. *International Journal on Green Economics*. 8(1): 19–35.

Niti Aayog. 2020. India 2020. IEA In-depth Energy Policy Review. Government of India, New Delhi.

Planning Commission. 2015. Government of India, New Delhi.

Pradhan Mantri Sahaj Bijli Har Ghar Yogana (SAUBHAGYA). 2018. Ministry of Power, Government of India, New Delhi.

Pulighe, G., G. Bonati, M. Colangeli, M. M. Morese, et al. 2019. Ongoing and emerging issues for sustainable bioenergy production on marginal lands in the Mediterranean regions. *Renewable and Sustainable Energy Reviews*. 103: 58–70

Rahman, S., N. A. Rahim, M. R. Islam and K.H. Solangi. 2011. Environmental impact of wind energy. *Renewable and Sustainable Energy Reviews*. 15(5): 2423–2430.

Rao, P. V. and S. S. Baral. 2013. Biogas in India. *Renewable Energy Akshay Urja, GoI.* 6(5 and 6): 48–51.

Ravindranath, N.H., H. L. Dattaprasad and H. I. Somashekar. 1990. A rural energy system based on energy forest and wood gasifier. *Current Science* 59(11): 557–560.

Ravindranath, N. H., H. I. Somashekar, S. Dasappa and C. N. Jayasheela Reddy. 2004. Sustainable biomass power for rural India: Case study of biomass gasifier for village electrification. *Current Science* 87(7): 932–941.

Reid, D. 2017. BBC News May 2017. https://www.bbc.com/news/av/technology-39963455/kamuthi-the-world-s-largest-solar-power-project

Singh, K. 2019. India's bioenergy policy. *Energy, Ecology and Environment.* 4: 253–260.

State of India's Environment. 2020. Center for Science and Environment, New Delhi.

The Economic Times. 2019. India ranks 5th in Global Climate Risk Index. https://economictimes.indiatimes.com/home/environment/global-warming/india-ranks-5th-in-global-climate-risk-index/articleshow/72367505.cms

The Guardian. 2016. India plans nearly 60% of electricity capacity from non-fossil fuels by 2027. https://www.theguardian.com/world/2016/dec/21/india-renewable-energy-paris-climate-summit-target

Welch, J. B. and A. Venkateswaran. 2009. The dual sustainability of wind energy. *Renewable and Sustainable Energy Reviews.* 13: 1121–1126.

World Bank, 2000. Development in practice, rural energy development, improving energy supplies for two billion people. Entering the 21st century, World Development Report 1999/2000, Oxford University Press.

6 Energy and Environmental Development in Maldives

Fathmath Shadiya

CONTENTS

6.1 The Maldives Overview.. 123
6.2 Development of the Energy Sector in the Maldives 124
    6.2.1 Energy in the Transport Sector of the Maldives............................. 126
    6.2.2 Energy in the Tourism Sector of the Maldives 127
    6.2.3 The Necessity for Energy Security in the Maldives........................ 128
6.3 The Need for Sustainable Development in the Energy Sector
    of the Maldives 129
    6.3.1 Impact of Climate Change on the Development of the Maldives..... 130
6.4 Current Trends and Future Predictions of the Energy Sector in the
    Maldives... 132
    6.4.1 The Future of Hybrid Energy Systems in the Maldives 135
    6.4.2 Challenges and Obstacles in the Evolving Energy Sector of the
    Maldives.. 136
6.5 Conclusion 137
References... 138

6.1 THE MALDIVES OVERVIEW

The Republic of Maldives is an island nation made up of 1192 low-lying coral islands. The pristine coral island nation is located 750 km south-west of Sri Lanka and stretches 115,300 km² north to south, across the Equator (Ministry of Environment and Energy 2012). There are 26 Atolls in the Maldives. Within these 26 Atolls, 187 islands are inhabited and 1005 islands are uninhabited. The total population of the Maldives in 2017 was 491,589 (National Bureau of Statistics 2018). The Maldives has achieved considerable economic and social progress since 1970 due to a successful tourism industry. Today, the tourism industry is the largest contributor to the Maldivian economy. Tourism accounts for about 30% of the gross domestic product (GDP) of the country. Apart from the tourism industry, the fisheries, commerce and construction industry also contributes to the economic growth of the Maldives. In 2017, the GDP of the country reached 6.9% and in the first quarters of 2018, the GDP growth reached an average of 9.1% (International Monetary Fund 2019). With rapid economic growth, the demand for energy is growing quickly across the country. Today, the energy sector is the fastest-growing sector in the Maldives.

TABLE 6.1

The Type of Fossil Fuel Imported to the Maldives for Different Purposes

Type of Fuel	Metric Tons	% Share
Cooking gas	14,483	3
Diesel	447,555	80
Petrol	57,730	10
Aviation gas	41,666	7
Total	561,434	100

Source: Ministry of Environment and Energy 2017b.

Historically, the main source of energy was biomass; however, as the Maldives progress economically, more and more woodlands got converted to urban areas, causing a reduction in the available biomass across the country. To compensate for the depleting biomass, Liquified Petroleum Gas (LPG) was introduced to the country as a domestic fuel. According to Table 6.1, by 2017, about 14,483 metric tons of cooking gas were imported. This is about 3% of the total fossil fuel imported into the country. Since diesel is used to generate electricity, diesel is the most dominant fossil fuel used in the Maldives. Diesel shares 80% of the fossil fuel market. Next to diesel, petrol is the second most consumed fossil fuel in the Maldives. Petrol shares 10% of the fuel market. Petrol is used mainly for inland transport vehicles (Ministry of Environment and Energy 2017b).

With new economic sectors developing rapidly across the Maldives, the demand for energy is increasing quickly. However, for a country such as the Maldives, where fossil fuel is 100% imported, increase in energy means the government has to spend a considerable amount of money to purchase fossil fuel from the international oil market. The Maldives is very vulnerable to external shocks in the international oil market; therefore, to ensure the sustainability of the country's economic development, the Maldives needs to move toward renewable energy technologies. But can the Maldives achieve energy security by adopting renewable energy technology? To answer this question, in the coming sections, first, the growth of the Maldivian energy sector will be described, next, the need for sustainable energy alternatives will be discussed, followed by a brief examination into the feasibility of renewable energy projects conducted in the Maldives, before concluding the chapter.

6.2 DEVELOPMENT OF THE ENERGY SECTOR IN THE MALDIVES

The Maldives was able to produce electricity for the first time in 1949, using a generator set of 6 kW in the capital city, Malé. Initially, electricity was supplied to the presidential palace and a few government office buildings. Later this was extended to the entire population of Malé. By 1997, the government formed a company called the

State Trading Electric Company (STELCO) to provide electricity to large populated islands of the Maldives as a means to eradicate extreme poverty. During that time, in most islands, electricity was generated using small diesel generators run by the local communities. These small community-based power systems were able to produce up to 12 hours of electricity, but with frequent power failures. The community-based power systems were not run efficiently due to lack of proper infrastructure, finance and technical experts available in the island's communities. Provision of electricity is essential for poverty alleviation; therefore, during the early 1990s, to improve the living standards of the most populated island communities, STELCO built 32 electricity facilities to provide 24 hours of uninterrupted electricity to these island communities (State Electronic Company Limited 2020).

By 2009, to upgrade the locally operated powerhouses, six new utility companies were formed in six different regions of the Maldives by the government. The main purpose of forming these new utility companies was to expand the energy sector of the Maldives to provide quality affordable utility services to all the inhabited islands of the Maldives (State Electronic Company Limited 2020). To make the energy sector even stronger, in 2012, the six utility companies were merged into one large company called FENEKA Corporation Limited, which was established on 18th June, 2012 to provide a secure and reliable supply of electricity, water, and waste management services to all the inhabited islands of the Maldives. Today FENEKA Corporation has 148 power-houses across 152 islands of the Maldives. FENEKA replaced most of the island community-based power systems in the Maldives. Table 6.2 summarizes the total number of powerhouses present in the Maldives (State Electronic Company Limited 2020).

With improved electricity services across the Maldives, the manufacturing industry, the transport sector and the tourism industry were able to expand their businesses throughout the country. Today, the manufacturing industry such as the fish canning industry uses sophisticated pieces of machinery to produce export quality canned tuna. Large-scale electrical equipment utilized in the trade industries requires a great deal of energy to operate (Abdulla 2014). Such as the manufacturing industry, the transport sector also consumes a lot of energy. Next to electricity generation, the transport sector is the second largest energy-consuming sector in the Maldives.

TABLE 6.2

Number of Powerhouses in the Maldives

Powerhouse Establishments	Number
FENEKA	148
STELCO	35
Malé Water and Sewerage Company (MWSC)	1
ISLAND COUNCILS	4
PRIVATE COMPANY	1
Total	189

Source: State Electronic Company Limited 2020.

6.2.1 ENERGY IN THE TRANSPORT SECTOR OF THE MALDIVES

In the past, Maldivians used sailboats to travel from one island to another; however, with the introduction of mechanized boats during 1970, people started to prefer mechanized boats over sailboats. Over the years, diesel-based engine boats completely replaced all the sailboats (Aboobakuru 2014). According to Table 6.3, Dhoni, a diesel-based engine boat is the most frequent type of vessel found in the Maldives, followed by speed boats. With the growth of local tourism in the Maldives, the demand for speed boats has been increasing quickly. Today speed boats account for 21% of the total transport vessels in the Maldives (Abdulla 2014).

Nowadays, not only diesel engines have replaced sailboats, but they have also become larger in size. The large boats with efficient engines can travel very far, which was not possible a few decades ago. In 2009, then president, Mohamed Nasheed introduced a nationwide inter-Atoll ferry system, to connect all the Atolls of the Maldives. The nationwide ferry system made transport between Atolls possible at an affordable rate for the locals. Maldives Transport and Contracting Company (MTCC) took the responsibility of providing the inter-Atoll ferry service to the public. MTCC is a state-owned company established in 1980 to develop the transport sector of the Maldives. At present, MTCC is the longest-serving public company in the country (Maldives Transport and Contracting Company 2019).

A low cost, affordable, transport system is vital for the economic and social progress of the Maldives. To make transport cheaper and to reduce the heavy dependency on diesel, the maritime sector must invest in cheaper renewable energy alternatives. However, the maritime sector is also challenged with a lack of available funding opportunities to build the adequate infrastructure needed for the installation of renewable energy technologies in the dockyards as well as transport vessels. Hence, it is essential to source finance to interested private parties to invest in renewable energy technologies. Besides, a stronger legal framework is also needed to attract private investors who want to invest in renewable energy technologies so that future of the maritime sector can be more secured (Abdulla 2014).

TABLE 6.3
The Percentage Share of Sea Transport Vessels Used in the Maldives in 2014

Vessel Type	% Share
Dhoni	57
Speed Boats	21
Boat	5
Yacht	1
Other	16

Source: Abdulla 2014.

Parallel to sea transport, land transport has changed over the years. Today, motorcycles, cars, lorries and trucks are the main means of land transport. In 2012, for every 3.5 persons living in Malé, there was one motorbike, and for every 50 people, there was one car (Abdulla 2014). At present, the transport sector releases 25% of greenhouse gas (GHG) emissions in the Maldives. With the increased usage of vehicles, traffic congestion, air pollution and noise pollution have become urban worries in the greater Malé region. Most of these vehicles are old and inefficient vehicles that release a lot of harmful gases to the atmosphere. To prevent degradation of the urban environment, mitigation of GHGs from the transport sector is needed. GHG emission from the transport sector is the second largest contributor to atmospheric pollution in the Maldives. It is alarming to know that the total registered vehicles have increased more than 295% between 2007 and 2014. About 83% of the registered vehicles are motorcycles, followed by cars (Abdulla 2014). Furthermore, the growing demand for energy means more dollars have to be spent by the government to purchase oil, which will increase the outflow of foreign exchange. Being a small island developing nation with limited industry, the outflow of foreign exchange will slow down the economic growth of the country very much (Ministry of Environment and Energy 2015).

6.2.2 Energy in the Tourism Sector of the Maldives

Next to the transport sector, the tourism industry is also a massive industry that consumes a lot of electricity in the Maldives. Tourism has become the single largest business sector in the world economy. Today, the tourism sector represents 10% of the global GDP. Globally, 1 out of every 10 jobs is related to the tourism industry (United Nations Development Programme 2018). The tourism sector is, directly and indirectly, related to sustainable development. This is indicated in the sustainable development goals (SDGs) adopted in 2015. There is a total of 17 goals with 169 associated targets in SDG. While all the goals in SDGs are directly and indirectly related to the tourism sector, goal 12 and goal 13 specifically have a massive influence in reducing excessive energy consumption in the tourism sector. Goal 12 is about *"Responsible consumption and protection"* which promote energy conservation. Under this goal, the tourism sector can encourage its stakeholders to become more responsible toward energy consumption by adopting energy-efficient technologies. At the same time, goal 13, which is *"Taking urgent action to combat climate change and its impacts"* encourages the tourism sector to fight against climate change by reducing the carbon footprint. Carbon footprint can be reduced by incorporating low-carbon technologies into the tourism sector (World Tourism Organization 2015).

Tourism in the Maldives began in 1972. Since then, the tourism industry has grown rapidly beyond adventure tourism and became more high-class oriented. In 1980, the number of tourists who came to the Maldives were 42,000 tourists and this number grew close to 800,000 by 2010 (Ministry of Tourism, Arts and Culture 2012). To ensure the success of the tourism industry, the government of the Maldives developed the first Tourism Master Plan in 1982. Strict legislation was developed in this Master Plan to ensure sustainable development from the very early stage of tourism development in the Maldives. The second Tourism Master Plan was developed in 1996 to guide the tourism industry to effectively involve the private sector in the development of the tourism industry. The third Tourism Master Plan was developed in 2007 to encourage sustainable development, especially in the energy sector. As each resort is housed with its diesel generator to produce electricity, the generator releases dangerous levels of carbon dioxide to the atmosphere. Hence, the new vision in the tourism sector is to lessen the reliance on diesel generators by working toward establishing renewable sources of energy in the resorts (Ministry of Tourism Arts and Culture 2012).

The third Tourism Master Plan was very much in alignment with the 7th National Development Plan of the Maldives. The 7th National Development Plan encourages the introduction of renewable energy technologies such as biofuel, solar and wind energy across the Maldives (Ministry of Planning and National Development 2007). It is interesting to note that even though the importance of renewable energy technology was mentioned in the 7th National Development Plan since 2007, it was during President Mohamed Nasheed's administration in 2009, renewable energy became a prominent subject in the Maldives when the president declared a 50% reduction in the usage of fossil fuel to generate electricity by 2020 (Ministry of Housing and Environment 2010). With the rising price of diesel, Maldives is always subjected to external shocks in the oil market, which leads to uncertainty in the future of the energy sector. Hence, energy insecurity is a central problem to the development of the Maldives.

6.2.3 THE NECESSITY FOR ENERGY SECURITY IN THE MALDIVES

Energy insecurity is not an exclusive problem to the Maldives only, looking at other small island states, similar issues can be observed. Such as the Maldives, smallest island states depend on expensive fossil fuels to generate electricity. Electricity is vital for the effective delivery of social services such as health, education, water and sanitation; however, affordable cheap electricity is still a challenge in most of the small island states. Currently, 70% of the Pacific residents do not have access to electricity and depend on wood, kerosene and batteries for energy (Roper 2005). Besides fuel-related impacts, small islands tend to pay more for electricity due to their small scale. Often due to geographic isolation, small island states do not have a single grid connection to large generators, as a result, the governments of small islands cannot provide benefits such as customer discounts or fuel subsidy to their citizens due to low return on investment (Stuart 2006).

In small island developing states, there is an intrinsic trade-off between achieving economy of scale and ensuring a secure supply of energy. Because electricity cannot be readily stored in small islands due to lack of space, supply must be continuously adjusted to varying demand, which is a challenging task due to inadequate stock. Capacity management is another key factor that needs to be considered by utility companies when developing energy projects in small island states. To increase capacity, it will be easy to purchase new generators with greater power. However, if modifications are needed to the already existing infrastructures, it will be nearly impossible to bring alterations due to the high cost involved in modification (Roper 2005). Therefore, to ensure the sustainability of energy projects, it is important to anticipate problems specific to a small island setting at the beginning of the projects.

6.3 THE NEED FOR SUSTAINABLE DEVELOPMENT IN THE ENERGY SECTOR OF THE MALDIVES

In 2008, the government started to provide 24 hours of uninterrupted electricity to all the inhabited islands, commercial and tourist resorts of the Maldives. But due to geographic isolation of the islands, the government could not provide electricity using a single grid. Instead, in each island, a mini power grid was installed to cater to the energy demands of that particular island. Today, with the increase in urbanization across the Maldives, the demand for electricity is growing rapidly. The demand is expected to rise exponentially in the greater Malé region, where one-third of the population resides (Ministry of Environment and Energy 2018). The greater Malé region is the most populated in the Maldives and consumes about 400 gigawatt hours (GWh) of electricity per year. The huge demand for electricity in this region is due to a greater influx of urban population seeking employment, health care and education. At present, it is estimated that the greater Malé region accounts for 56.9% of the total electricity produced in the Maldives compared to the rest of the country. And this figure is expected to increase at a rate of 8.5% per year.

In the Atolls, due to being less populated, electricity consumption is low compared to the greater Malé region; yet, provision of electricity is still expensive for the government, as the government bears all the expenses needed to operate the powerhouses. Furthermore, the high production cost of electricity necessitates the government to give subsidy to the state-owned utility companies so that they can provide electricity at an affordable price to the general public. The government provides an electric subsidy to the utility companies to maintain the stability of energy prices at an affordable price for the public. However, the electricity subsidy has imposed a significant burden on the government's budget. In 2011, the government has to spend USD\$25 million on electricity subsidy alone, and this figure is expected to rise with an increase in electricity usage (Ministry of Environment and Energy 2017a). When the government is required to spend an immense amount of money to provide affordable electricity, it will be challenging for it to invest in additional development projects the country urgently needs.

The term "energy insecurity" has evolved over the years. Until the oil crisis in the 1970s, energy insecurity was mostly linked to the geopolitical risks connected with securing energy supplies. But over the past few years, energy insecurity and sustainable development have been at the forefront of the global agenda due to the high impact of volatile energy prices on the lives of people and their surroundings. To secure energy, countries must ensure they receive a sufficient supply of energy at an affordable price in a sustainable manner. This means taking serious actions against factors that influence energy insecurity such as lack of available diverse energy resources, high import dependency, absence of secure safe transit routes, and shortage of reliable infrastructure to store energy supplies (Shumais and Mohamed 2019). Tackling these factors needs to be a government priority to protect the future of the energy sector.

Energy has a supply and a demand side. To ensure energy security, both the supply and the demand sides must be considered equally. The supply side of energy security includes the ability of a country to generate and produce energy efficiently from varieties of sources, while at the same time effectively managing energy sources available within and outside of the country. The demand side of energy security requires access to energy services at an affordable price by the general public (Shumais and Mohamed 2019). Energy insecurity can affect the socio-economic development of a country, as energy insecurity can have a direct impact on the livelihood of the people when fuel price increases. An increase in fuel price can push people into poverty, as they have to choose between paying for energy or buying essentials such as food (Shumais and Mohamed 2019). For example, in many developing parts of the world, if fuel prices become expensive, powerhouses, very often cut off electricity to reduce fuel costs. This practice affects people's ability to work. For poor people, the inability to work will worsen their already poor circumstances (United Nations Development Programme 2007)

Energy insecurity will continue to exist in the Maldives as long as the Maldives depends on fossil fuel for energy. Therefore, a long-term commitment to improving resource efficiency in the energy sector is important to overcome existing barriers in the energy sector (United Nations Development Programme 2007). To safeguard the Maldives from energy insecurity, the Ministry of Environment has developed an energy policy called "The Maldives Energy and Strategy 2010". The main aim of this policy is to provide reliable and sustainable energy resources to the local community. The policy was revised in 2016 with new key priority themes, such as: (1) Affordable electricity to all citizens, (2) Energy security using renewable energy resources, and (3) Energy conservation (Ministry of Environment and Energy 2016). Attaining clean technology is crucial for the Maldives to reduce the high dependency on fossil fuels and to reduce air pollution.

6.3.1 IMPACT OF CLIMATE CHANGE ON THE DEVELOPMENT OF THE MALDIVES

Climate change and sea-level rise will have a dramatic impact on coastal communities such as the Maldives. With an increase in intense destructive weather systems, livelihood activities such as agriculture, fisheries and tourism will suffer immensely. Change in climate conditions such as sea-level rise, frequent tropical cyclones, storm surges, and shoreline erosions will threaten the sustainability of the tourism industry unless serious mitigation and adaptation measures are taken quickly (Ministry of Environment 2017).

In the face of climate change, there is not much small island states such as the Maldives can do; however, if the Maldives become secure in energy, the country's economy can be protected from international turmoils often experienced in the energy market. Goal 7 of the sustainable development goals (SDGs) is about ensuring access to affordable, reliable modern energy to all the citizens of the globe. The SDGs require integration of economic, social, cultural, political and ecological factors to ensure a better quality of life (Douglas 1997). The principles of sustainability depend on equal distribution of existing resources. Therefore, governments should have a practical perspective when developing policies to eradicate poverty by not excluding the most vulnerable in society. However, in the case of small island states, development is mainly concentrated only in a few islands known as the core centres. As a result, a large percentage of the rural communities who are excluded from the benefits of modern development often migrate into these core centres to pursue employment, education and a better quality of life. Every year, as more people settle in these core centres, electricity demands in the core centres increase rapidly (Douglas 1997).

Energy offers the opportunity for improved livelihood and economic progress. Countries cannot reduce poverty without investing in energy. Provision of energy can be grouped into (1) energy accessibility, (2) energy efficiency, and (3) share of renewable energy (Yoshida and Zusman 2014). Energy accessibility can be correlated to the level of development that the country is experiencing. Thus, poor countries need to prioritize accessibility of energy as an essential component needed for the well-being of society. However, it is very unlikely that poor countries can invest heavily in the energy sector unless developed countries provide the financial and technical assistance needed by these countries. Many parts of the world do not have access to affordable energy. Provision of clean energy can contribute to wider social development and economic opportunities, while at the same time fighting against climate change.

In the Maldives, the Ministry of Environment is responsible for formulating policies related to energy. Out of the six departments in the Ministry, the energy department is accountable to develop strategies needed to address the growing needs in the energy sector of the Maldives. The energy department is divided into two sections: the *"Policy and Sector Development Section"* and the *"Energy and Technology Development Section"*. The *"Policy and Sector Development Section"* is responsible for developing policies needed to expand the energy sector of the Maldives, while the *"Energy and Technology Development Section"* is responsible for introducing innovative technologies to overcome the misuse and wastage of energy in the country (Ministry of Environment 2019). At present, the government of the Maldives is aiming to scale up renewable energy installation from 2 MW to 21 MW (Ministry of Environment and Energy 2012).

Renewable energy technology is widely spreading across the world, and the prices of renewable energy technology are falling, creating new possibilities of affordable energy services across the globe (The World Bank 2019). In the Maldives, the government is participating in several renewable energy projects. One such notable project is called the "Scaling Renewable Energy Program (SREP)". The Maldives is one of the six pilot countries participating in this project. The main objective of the project was to transform the contemporary fossil fuel-based energy systems into modern renewable energy systems. Investing in renewable energy systems will benefit the Maldives, as these investments will reduce the country's heavy dependency on fossil fuels (Ministry of Environment and Energy 2012).

Though the government of the Maldives is investing in renewable energy technologies, it is worthy to mention that, without foreign support, it will not be easy for the government to transform the current conventional energy systems to modern renewable energy systems due to high costs involved in such projects. For a country such as the Maldives where resources are limited, financial and technical assistance will be needed to implement expensive projects effectively. Often, when investment costs are too high, developing countries are forced to choose cheaper alternatives to provide electricity at an affordable price to the citizens (Shumais and Mohamed 2019). Energy accessibility is a top priority of the government, and unless the cost of renewable energy is economical, it will not be feasible for the government to shift from fossil fuel-based energy systems to renewable energy systems completely.

Even so, the Maldivian government is putting effort to transmute the conventional energy systems in the country to renewable energy systems by exploring cheaper indigenous energy resources available in the country. The government hopes to produce a minimum of 30% of daytime peak load of electricity using renewable energy sources, and to attain this goal, guiding principles were developed in the revised energy policy "*Maldives Energy Policy and Strategy 2016*". The energy policy encourages the government and the private sector to explore indigenous energy resources to diversify the available energy resources in the Maldives. The guiding principles also encourage the public to embrace healthy lifestyles as a means to adopt energy-efficient technologies to reduce carbon emission (Ministry of Environment and Energy 2016).

Installing renewable energy systems in a country such as the Maldives comes with difficulties, due to reasons such as the absence of essential data about the accessible renewable energy resources in the Maldives, lack of evidence on the feasibility of installing new energy options, shortage of technical capacity to design, develop and install novel energy technologies, and absence of financing mechanisms for the private and public companies involved in clean energy technologies (United Nations Development Programme 2007). Even though there are many difficulties to overcome in establishing an effective renewable energy sector in the Maldives, successful installation of renewable energy systems will result in cleaner, safer and cheaper energy opportunities for the Maldivians, which is essential to combat climate change. At the same time, renewable energy systems will protect the Maldivian economy from fluctuating oil prices in the international oil market, which is needed to safeguard the Maldivian economy in the long run (Shumais and Mohamed 2019).

6.4 CURRENT TRENDS AND FUTURE PREDICTIONS OF THE ENERGY SECTOR IN THE MALDIVES

The Paris Agreement brings all the nations into a shared cause to combat against climate change. The Paris Agreement entered into force on 4th November, 2016. The main aim of the Paris Agreement is to strengthen the global response to the threat of climate change. The Maldives signed the Paris Agreement on 22nd April, 2016 (United Nations 2020). Upon signing this agreement, achieving carbon neutrality became a top government priority. Hence, to encourage renewable energy technology in the Maldives, the government developed several policy documents. A summary of these documents is given in Table 6.4.

The foremost objective of the policy documents given in Table 6.4 is to encourage the Maldivian government to espouse renewable energy technologies as a means to ensure the sustainability of the energy sector. However, all the policy documents given in Table 6.4 lack clear measurable targets to achieve this goal. The targets given in the policy documents were descriptive. Descriptive targets often lead to ambiguity of the desired results. Energy insecurity will continue to exist in the Maldives as long as fossil fuel dominates the energy sector; therefore, clear-cut measurable targets are vital to be incorporated into the future policies to reduce the uncertainty of the desired results, especially in a time when people's behavior is changing due to improvement in living standards.

TABLE 6.4
List of Policy Documents and Action Plans Developed to Encourage Carbon Neutrality in the Maldives

Strategy Basis	Year Developed	Strategy Summary
1. The third National Environment Action Plan (NEAP)	2008	NEAP details the plan for environmental protection and management from 2009–2013, including a more detailed approach for the carbon neutrality goal, increasing the resilience of islands and capacity development plans (Ministry of Housing, Transport and Environment 2008)
2. The Strategic Action Plan (SAP)	2019	Includes all broad activities for mitigation, adaptation and increasing resilience, as well as governance reforms needed (The President Office 2019)
3. Maldives National Strategy for Sustainable Development (NSSD)	2009	Looks at carbon neutrality as well as the protection of islands for sustainable development and considers food security (United Nations Environment Programme 2009)
4. Maldives National Energy Policy and Strategy	2010	Focuses on achieving carbon neutrality, energy conservation and efficiency to secure the energy sector (Ministry of Environment 2020)
5. Maldives Scaling Up Renewable Energy Program (SREP) Investment Plan	2012	Aims to encourage private sector involvement in the renewable energy sector of the Maldives (Ministry of Environment 2020)
6. The Maldives Low Carbon Strategy	2014	Presents an overview across sectors of potential mitigation of Greenhouse Gas—emissions attainable for the Maldives in the short term and the social costs attached with such mitigation (Ministry of Environment 2020)
7. Low Carbon Strategy for the Transport Sector	2014	Provides a background analysis of the impacts of climate change from the transport sectors, and it addresses the potential strategies the transport sector can adopt to reduce GHGs (Ministry of Environment 2020)

In the Maldives, over the years, as the living standard of people changes due to improvements in economic progress, the demand for electrical appliances such as air conditioners has also increased throughout the country. Since 2012, the demand for air conditioners has been increasing rapidly, as shown in Figure 6.1. Today, air conditioners account for 60% to 70% of electricity consumption in households and offices (Shumais and Mohamed 2019). The demand for electrical appliances is not exclusive to the greater Malé region only. With the introduction of local tourism in 2009, the demand for electrical appliances in the outer islands also has been increasing rapidly. However, unlike the capital city Malé, the outer islands have small generators with low capacity; therefore, during the peak time when electricity usage

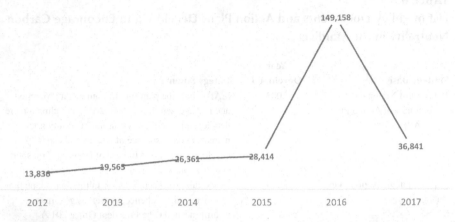

QUANTITY OF AIR CONDITIONERS IMPORTED TO THE MALDIVES

149,158

36,841

26,361 ———— 28,414

19,565

13,838

2012   2013   2014   2015   2016   2017

FIGURE 6.1 Quantity of air conditioners imported to the Maldives (Source: Shumais and Mohamed 2019).

increases, frequent power failures occur. As many tourists show their dissatisfaction with electricity shortages, recurring power failures can threaten the budding local tourism in the islands. Hence, to protect the future of local tourism, the local islanders need to adopt energy-efficient approaches or they may need to capitalize on generators with higher capacities to prevent power failures. Since both possibilities are not easy options to implement, the future of local tourism is very vulnerable, which in turn may impact the livelihood of people who depend on local tourism for income (Ministry of Environment and Energy 2016).

The Maldives has an abundance of sunshine, and investing in solar photovoltaic (PV) is highly sensible from the perspective of energy security. However, the lack of a vast area of land on the islands will be challenging when it comes to installing solar PVs. As the islands are small, one of the difficulties will be to find appropriate spaces to install solar panels with continuous sunshine. There are cases in which tree removal has taken place over a relatively large area of islands for installation of solar panels (Shumais and Mohamed 2019). Where possible, it is necessary to minimize this, as the conservation of trees is important to maintain terrestrial biodiversity of the Maldives. Hence, the Maldives could try to incorporate diverse types of technologies or explore new alternatives that will require less space. An alternative is to install floating solar PV panels on lagoons, close to the shore. However, this process could also be challenging, as lagoons are places where most islanders dock their boats. An option worth exploring will be hybrid energy systems. Research conducted in the Maldives shows 100% renewable energy systems are not financially viable due to high investment costs and lack of land. On the other hand, hybrid systems which can run on existing diesel generators, in combination with PV solar panels and electricity storage have shown the potential of lowering the price of electricity from 0.05 to 0.10 US\$/kWh (Wijayatunga et al. 2016).

Achieving carbon neutrality in the Maldives can be a lengthy process due to many phases involved in the process, such as securing of investments opportunities, capacity buildings to operate and manage renewable energy systems and public awareness. Each of these phases will take time, effort and money to execute (Ministry of Environment 2007). While there is no doubt that renewable energy technology is needed to secure the future of the energy sector of the Maldives, attention must also be given to energy conservation. Currently, in the Maldives, renewable energy technology such as PVs is only applied in navigation lights, in outer island communication facilities and some resorts for heating water. However, there are many opportunities for renewable energy applications in the Maldives. Some of these possibilities are hybrid systems using wind or solar with diesel generators or landfill gas, etc. to produce electricity for multiple uses (Wijayatunga et al. 2016).

6.4.1 THE FUTURE OF HYBRID ENERGY SYSTEMS IN THE MALDIVES

Hybrid power generator systems are economically attractive and environmentally appealing. Successful implementation of such a system can reduce the electrical cost for the government. At present, the Ministry of Environment in collaboration with the Asian Development Bank (ADB) is participating to test the feasibility of solar battery hybrid diesel system under the project titled, Preparing Outer Islands for Sustainable Energy Development (POISED). The main aim of the project is to transform the existing diesel-based mini-grids into hybrid renewable energy systems across the Maldives. The project aims to construct and operate hybrid energy systems with solar power generation units of 20888 kWp capacity and 40170 kW diesel generations in 160 selected islands (Ministry of Environment and Energy 2017b). However, the success of the project is yet to be determined as hybrid systems also have challenges to overcome.

To ensure the success of hybrid energy systems, one of the biggest challenges the government must overcome will be to retain trained staff who know how to operate and manage hybrid systems in the islands. Often, trained people migrate to the capital city Malé to seek better employment, health or education opportunities. Staff retention is important because, without trained staff, the Maldives will continue to depend on foreign relief for technical support, which is often expensive and time-consuming (Shumais and Mohamed 2019). Also, the government must overcome the slow acceptance of new technology by the local communities. Installing renewable energy technology is often challenging at the island level, as most islanders hesitate to adopt unfamiliar sources of energy when they are so used to the familiarity of fossil fuel-based energy systems. Often, island communities prefer to take a wait-and-see approach rather than embracing new methods (Berkhout and Marcotullio 2012).

To encourage the local communities to embrace new technologies, incentives such as financing schemes need to be introduced. In the Maldives, financing schemes considering energy efficiency include the Bank of Maldives "Green Loan scheme" and the "Fund for Renewable Energy System Applications (FRESA)" scheme. Apart from providing finance to energy-efficient schemes, the FRESA scheme also provides investment opportunities to develop solar PV systems, windpower systems, bio-digesters projects, solar technology, etc. However, due to lack of community awareness and the high cost of initial investments, the demand for these financial schemes is very low in the Maldives (Bank of Maldives 2016). Designing financial incentive schemes will not be easy. There are many things to consider when designing a financial scheme. Market studies need to be carried out to select if the technologies included in the finance scheme have a good prospect in the future once the scheme is released. Carefully thought incentives strategically needs to be in place to encourage people to adopt new technologies. It is also very important to leverage available incentive funds and set incentives to match incremental costs on a life-cycle basis. Rewarding performance to minimize free riders is also an essential component of a financial scheme (Loorbach and Verbong 2012).

Despite existing financial schemes and projects related to renewable energy, the transition from fossil fuel-based energy systems to renewable energy systems can be a very slow process. The energy transition can be defined as "a shift from the current energy usage which relies primarily on non-renewable energy resources to a more efficient lower carbon energy resource" (Berkhout and Marcotullio 2012). Scholars of transition studies explain that, for a successful conversion to take place, a combination of economic, political, institutional and socio-cultural adaptation strategies will be required by the government. This is because price alone does not guarantee the successful implementation of new technologies. Energy markets are highly influenced by national and international policies and the key decision-makers regarding energy policies are the governments. Therefore, unless the government formulates strategies to create enabling conditions for new technology to emerge, the process of transition will be very slow to take place (Berkhout and Marcotullio 2012).

6.4.2 CHALLENGES AND OBSTACLES IN THE EVOLVING ENERGY SECTOR OF THE MALDIVES

In the context of the Maldives, electricity savings are not nearly fully utilized despite their financial and economic benefits. There is an expectation that the electricity demand in the Maldives will grow by more than 10% per annum unless the Maldives heavily invests in renewable energy systems (Abdulla 2014). For the Maldives to reduce carbon emission and to lessen the dependency on fossil fuel, hybrid energy systems look very promising, as these systems are cheaper and much easier to implement than renewable energy systems (The World Bank 2019). Even for hybrid systems, authorities need to review the complete energy chain, from the source of supply to ultimate consumption and waste. Besides, authorities also must find the application of hybrid systems to solve energy problems that exist in transportation, urbanization, and built environment to facilitate energy transformation (Stuart 2006).

The greatest obstacles for the transition from fossil fuel-based energy systems to renewable energy systems perhaps could be the convenience of fossil fuel (Wijayatunga et al. 2016). However, at the same time, it is worthy to mention that, in the modern world, climate change is a strong driver to encourage countries to cut back on their usage of fossil fuel. The uncertainty surrounding climate change is driving countries to adopt renewable energy technologies to reduce carbon emission. However, both mature and emerging economies still largely demand energy-intensive sources and services, especially to generate electricity. Emerging economies are completely dependent on fossil fuel-based energy systems; in these economies, the convenience of fossil fuel-based energy systems remains deeply engrained in all the societal domains and practices. Therefore, the transition will be harder in emerging economies, compared to matured economies (Wolfley 2018). Transition is a complex process with a huge number of driving factors and impacts that involves markets, networks, institutions, technologies, policies, individual behavior and autonomous trends. The key message we are learning here is that even though renewable energy systems have many advantages, the transition from renewable energy to fossil fuel-based energy systems will take time to accomplish (Loorbach and Verbong 2012). What is certain is that even though there is no simple solution for an energy transition, a transition is inevitable to countries such as the Maldives where climate change and energy insecurity threatens the development of the Maldives.

6.5 CONCLUSION

The Maldives has achieved considerable economic and social progress since 1970 due to a successful tourism industry. Apart from the tourism industry, fisheries, commerce and construction industries also contribute to the economic growth of the Maldives. With new economic sectors developing rapidly across the Maldives, the demand for energy is increasing quickly. However, for a country such as the Maldives, where fossil fuel is 100% imported, increase in energy means the government has to spend a considerable amount of money to purchase fossil fuel from the international oil market. The Maldives is very vulnerable to external shocks in the international oil market; therefore, to ensure the sustainability of the country's economic development, the Maldives needs to move toward renewable energy technologies.

Energy insecurity is one of the biggest problems that threaten the sustainable development of the country. Energy insecurity is not an exclusive problem to the Maldives, looking at other small island states, similar issues can be observed. Such as the Maldives, most small island states depend on expensive fossil fuels to generate electricity. Electricity is vital for the effective delivery of social services such as health, education, water and sanitation; however, affordable electricity is still a challenge in most of the small island states. Energy insecurity can affect the socio-economic development of a country, as energy insecurity can have a direct impact on the livelihood of the people when fuel price increases. An increase in fuel price can push people into poverty as they have to choose between paying for energy or buying essentials.

Energy insecurity will continue to exist in the Maldives as long as the Maldives depends on fossil fuel for energy. Therefore, a long-term commitment to improving resource efficiency in the energy sector is important to overcome existing barriers. Also, climate change and sea-level rise will have a dramatic impact on coastal communities such as the Maldives. Though the government of the Maldives is investing in renewable energy technologies, it is worthy to mention that, without foreign support, it will not be easy for the government to transform the current conventional energy systems to modern renewable energy systems due to high costs involved in such projects. For a country such as the Maldives where resources are limited, financial and technical assistance will be needed to implement expensive projects effectively. Installing renewable energy systems comes with complications. Even so, successful installation of renewable energy systems will result in a cleaner, safer and cheaper energy opportunities for the Maldivians, while at the same time protecting the Maldivian economy from fluctuating oil prices in the international oil market, which is essential to safeguard the Maldivian economy in the long run. The Maldives has an abundance of sunshine, and investing in solar PV is highly sensible from the perspective of energy security. However, the lack of a vast area of land availability in the Maldives will be challenging when it comes to installing solar PVs. On the other hand, hybrid power systems are economically attractive and environmentally appealing. Even so, a transition from conventional energy systems to hybrid systems will be slow to take place in the Maldives. However, as the Maldives seek alternative energy resources to become less dependent on fossil fuel, a transition will be inevitable in the future of the Maldives.

REFERENCES

Abdulla, Adham Ahmed. 2014. *Low Carbon Strategy for the Transport Sector.* Government Report, Malé: Ministry of Environment and Energy.

Aboobakuru, Maimoona. 2014. "Transport Services in the Maldives- An Unmet Need for Health Service Delivery." *Transport and Communication Bulletin for Asia and the Pacific* 84.

Bank of Maldives. 2016. *BML launches Green Loan to encourage investment in environment-friendly technology.* February 28. Accessed November 15, 2017. https://www.bankofmaldives.com.mv.

Berkhout, Frans, and Peter Marcotullio. 2012. "Understanding Energy Transitions." *Sustainability Science* 109–111.

Douglas, C. H. 1997. "Sustainable Development in European Union States Small Island Dependencies- Strategies and Targets." *European Environment* 181–186.

International Monetary Fund. 2019. *Maldives 2019 Article IV Consultation-Press Release; Staff Report; and Statement by the Executive Director for Maldives.* IMF Country Report, Washington DC.

Loorbach, Derk, and Geert Verbong. 2012. *Governing the Energy Transition: Reality, Illusion or Necessity?.* New York: Routledge.

Maldives Transport and Contracting Company. 2019. *MTCC Homepage.* Accessed January 2020. https://mtcc.mv/.

Ministry of Environment. 2007. *Maldives: Renewable Energy Technology Development and Application Project.* Mid Term Review, Malé: Ministry of Environment.

Ministry of Environment. 2017a. *Voluntary National Review for the High Level Political Forum on Sustainable Development 2017.* Government, Malé: Ministry of Environment.

Ministry of Environment. 2019. *Sustainable Development Goals: Communication Strategy and Action Plan 2019-2023.* Government, Malé: Ministry of Environment.

Ministry of Environment. 2020. *Ministry of Environment: Republic of Maldives.* Accessed April 29th, 2020. http://www.environment.gov.mv/v2/en/types/annualreports.

Ministry of Environment and Energy. 2012. *Maldives SREP Investment Plan 2013-2017.* Investment Plan, Malé: Ministry of Environment and Energy.

Ministry of Environment and Energy. 2015. *Maldives Climate Change Policy Framework.* Policy Report, Malé: Ministry of Environment and Energy.

Ministry of Environment and Energy. 2016. *Maldives Energy Policy and Strategy.* Ministry of Environment and Energy

Ministry of Environment and Energy. 2017b. *Island Electricity Data Book 2017.* Data Book, Malé: Ministry of Environment and Energy.

Ministry of Environment and Energy. 2018. *Island Electricity Data Book 2018.* Government Report, Malé: Ministry of Environment and Energy.

Ministry of Housing and Environment. 2010. *Maldives National Energy Policy and Strategy.* Government Report, Malé: Ministry of Housing and Environment.

Ministry of Housing, Transport and Environment. 2008. *Third National Environment Action Plan.* Government, Malç: Ministry of Housing, Transport and Environment.

Ministry of Planning and National Development. 2007. *Seventh National Development Plan 2006-2010: creating new opportunities.* Government, Malé: Ministry of Planning and National Development.

Ministry of Tourism Arts and Culture. 2012. *Third Tourism Master Plan.* Government, Malé: Ministry of Tourism Arts and Culture.

National Bureau of Statistics. 2018. *Statistical Pocketbook of Maldives, 2018.* Report, Malé: Ministry of National Planning and Infrastructure.

Roper, Tom. 2005. *Small Island States-Setting an Example on Green Energy Use.* Report, Oxford: Blackwell Publishing Ltd.

Shumais, Mohamed, and Ibrahim Mohamed. 2019. *Dimensions of Energy Insecurity on Small Islands: The Case of the Maldives.* Report, Tokyo: Asian Development Bank Institute.

State Electronic Company Limited. 2020. *Our Company.* Accessed February 2020. https://www.stelco.com.mv/our-company.

Stuart, Kathy E. 2006. "Energizing the Island Community: A Review of Policy Standpoints for Energy in Small Island States and Territories." *Sustainable Development* 139–147.

The President Office. 2019. *SAP: Strategic Action Plan 2019-2023.* Government, Malé: The President Office.

The World Bank. 2019. *Energy Storage Roadmap for the Maldives.* Report, Washington DC: World Bank.

United Nations Development Programme. 2007. *Regional Energy Programme for Poverty Reduction UNDP Regional Centre in Bangkok.* Bangkok: UNDP Regional Centre.

United Nations Development Programme. 2018. *Tourism and the Sustainable Development Goals-Journey to 2030.* Report, United Nations Development Programme.

United Nations Environment Programme. 2009. *Maldives National Strategy for Sustainable Development.* Government, Malé: United Nations Environment Programme.

United Nations. 2020. *The Paris Agreement.* Accessed March 2018. https://unfccc.int/process-and-meetings/the-paris-agreement/the-paris-agreement.

Wijayatunga, Priyantha, Len George, Antonio Lopez, and Jose A Aguado. 2016. "Integrating Clean Energy in Small Island Power Systems: Maldives Experience." *Elsevier* 274–279.

Wolfley, Kathryn. 2018. "The Challenges of Climate Change to Energy Transitions Research." *Postgraduate Interdisciplinary Journal* 25–30.

World Tourism Organization. 2015. *Tourism and Sustainable Development Goals.* Report, Madrid: World Tourism Organization.

Yoshida, Tetsuro, and Eric Zusman. 2014. "Designing and Implementing an Energy Goal: Delivering Multi-benefits for Sustainable Development." *Institute for Global Environment Strategies.* Accessed March 2020. https://www.jstor.org/stable/resrep00750.

7 Energy Security in the Context of Nepal

Tri Ratna Bajracharya and Shova Darlamee

CONTENTS

7.1 Introduction 141
7.2 Energy Security in the Context of Nepal 144
7.3 Energy Security Assessment—A Case Study of Nepal 148
    7.3.1 Research Methods Adopted 148
    7.3.2 Results and Discussions 155
        7.3.2.1 Scenario Analysis 155
        7.3.2.2 Energy Security Indicators Formulation 161
        7.3.2.3 Principal Component Analysis 162
        7.3.2.4 Energy Security Index 163
        7.3.2.5 Environment Security 165
7.4 Conclusion 168
References 169

7.1 INTRODUCTION

Energy is an important component for fulfilling basic needs of people on day-to-day basis. Energy is also important for all economic activities which are necessary for production of goods and provision of services for any country. Any disturbances in the supply of energy could hamper the industrial or commercial growth of a country and thus could be detrimental to itseconomic development. Therefore, securing energy supply to meet national demand for both the short term and long term is one of the most important priorities for any country in the world, and some countries consider ensuring energy security as a national goal (Phdungsilp, 2010).

The concept of energy security basically relates to ensuring smooth energy supply in order to fulfill the energy demand. There has been a rapid growth in industrialization of developing countries causing growth in energy demand and consumption which in turn raises many issues such as frequent shortages in energy supply, increasing fuel prices, price volatility, environmental degradation and even geopolitical concerns. The concept of energy security is seen to be evolving over the years. While historically energy security studies were concerned with ensuring smooth supply of oil in particular, the concept of energy security evolved with ensuring supply of other energy sources. Furthermore, recent studies encompass not only ensuring supply, but also focusing on social, economic, environmental, military concerns related to energy supply and demand. Most of the literature on energy security focuses on energy risk, treating energy security and energy risk at the same time. Gonzalo et al. (2013) classified the overall energy risks as causes or factors capable of interrupting supply (primary energy risks) including geopolitical and technical factors, their effects (secondary energy risks) such as volatility of prices, interruption of supply, risks to human health and property and environmental risks and vulnerabilities. This gives the extent of exposure to risks.

In recent years, there have been various definitions of energy security suggested; for example, the IEA (2008) defines energy security as, "the uninterrupted availability of energy sources at an affordable price". The analysis of energy security in future scenarios uses the most generic definition of energy security as "low vulnerability of vital energy systems" (Cherp and Jewell, 2011). In a review by Ang et al. (2015), it is suggested that the meaning of energy security is highly context-dependent, such as a country's special circumstances, level of economic development, perceptions of risks, and the robustness of its energy system and prevailing geopolitical issues. Also, he added that the definitions and dimensions of energy security are dynamic and evolve as circumstances change over time. Some of the definitions of energy security found in the literature are shown in Table 7.1.

As there are a wide variety of energy security definitions available in the literature, there are also different approaches and methodologies toward both qualitative and quantitative assessment of energy security. Månsson et al. (2014) reviewed and suggested that energy security is a multidisciplinary rather than interdisciplinary concept with a variety of methodologies according to the researcher's background in multiple disciplines. These disciplines include:macro-economic using methodologies such as partial equilibrium models, cost benefit analysis, etc.; microeconomics using market behavior studies; industrial organization focusing on market risk exposure; financial theory using methodologies such as financial portfolios, real options theory, etc.; engineering research focused on reliability of power systems; operations research using multicriteria analysis, analytical hierarchy processes, etc.; political science focusing on international relations theory; complex system analysis applying simulation; dynamic system modeling methods; general system analysis using methodologies such as energy system scenario analysis and complex indicators and indexes, etc. Cherp and Jewell (2011) distinguished two most fundamental methodological choices in energy security assessments that are the choice between facts and perceptions in deciding what constitutes a significant energy security concern and concludes that the methodological choice should be systematic, rational and transparent.

There are many studies on energy security that often develop multidimensional metrics or indicators for conceptualizing energy security, or they measure energy security performance. The IEA (2007) devised a different set of metrics to evaluate the risk of system disruptions, imbalances between supply and demand, regulatory failures, and diversification among a subset of Organization for Economic Cooperation and Development(OECD) countries. The Energy Research Center of the Netherlands has also developed a comprehensive "Supply and Demand Index" to better assess diversification of energy sources, imports and suppliers, the long-term political stability in origins of supply, and rates of resource depletion (Scheepers et al., 2006). APERC (2007) used

TABLE 7.1
Definitions of Energy Security

Definitions	Sources
"The physical availability of supplies to satisfy demand at a given price."	(IEA, 2001)
"Energy security means ensuring countries can sustainably produce and use energy at reasonable cost in order to facilitate economic growth and, through this, poverty reduction; and directly improve the quality of peoples' lives by broadening access to modern energy services."	(World Bank, 2005)
"The ability of an economy to guarantee the availability of energy resource supply in a sustainable and timely manner with the energy price being at a level that will not adversely affect the economic performance of the economy."	(APERC, 2007)
"The energy security of developing countries refers to "enough energy supply (quantity and quality)" to meet all requirements at all time of all citizens in affordable and stable price, and it also leads to sustain economic performance and poverty alleviation, better quality of life without harming the environment."	(Martchamadol and Kumar, 2012)
"Equitably providing available, affordable, reliable, environmentally benign, proactively governed and socially acceptable energy services to end-usersinvariably fuses traditional conceptions of national security with emerging concepts of human rights, sustainable development, and individual security."	(Sovacool and Valentine, 2013), (Wang and Zhou, 2017)

the 4A framework for energy security assessment for the first time with dimensions: availability, affordability, acceptability and accessibility. It also used 5 energy security indicators: diversification of energy supply sources, net energy import dependency, non-carbon-based fuel portfolio, net oil import dependency, and Middle East oil import dependency, all under the 3 elements of physical energy: security, economic energy security and environmental sustainability. Vivoda (2009) sought to create a "novel methodological" approach to energy security and proposed 11 broad dimensions and 44 attributes that could be utilized to assess national performance on energy issues and used it to evaluate energy security in the Asia-Pacific region. Kruyt et al. (2009) proposed 24 simple and complex indicators for energy security. In an investigation of energy security, the Association of Southeast Asian Nations (ASEAN), under the 4A framework of availability, acceptability, affordability and applicability tracked the trends of each indicator for ten member countries in the years 2005–2010 (Tongsopit et al., 2016). A recent comprehensive literature survey by Ang et al. (2015) indicates diversity among studies that identify energy security indicators; the number of energy security dimensions ranging from 1 to 20 and the number of indicators ranging from 1 to 320. Chuang and Ma (2013) used a multidimensional criteria system consisting of dependence, vulnerability, affordability, and acceptability and six specified indicators to assess the effectiveness of Taiwan's energy policies on its energy security. Shin et al. (2013) simulated the effect of key policies on the improvements of 19 key security indicators based on quality function deployment and system dynamics. Yao and Chang (2014) used five metrics to analyze the trend of China's energy security over 30 years of reform. Martchamadol and Kumar (2012) developed the "AESPI" by combining 25 individual indicators in social, economic and environmental aspects to assess energy security of the past and future status. Wu et al. (2012) used 14 indicators to assess the relationship between climate protection and China's energy security. Sovacool and Ren (2014) suggested that energy security best consists of the four dimensions of availability, affordability, acceptability and accessibility using a DEMATEL (Fuzzy Decision-making Trial and Evaluation Laboratory) methodology to determine the most salient and meaningful dimensions and energy security strategies and apply the results to China, where they conclude that among the four energy security As; availability and affordability are more influential to the energy security metrics than acceptability and accessibility. The energy security of China was examined in a study by Yao and Chang (2014) using 4A framework of availability of energy sources, applicability of technology, societal acceptability and the affordability of energy resources and 25 indicators over 30 years of its economic reform showing that China needs to develop renewable energy resources on a large scale and control emissions. Sovacool (2013) constructed an energy security index to measure national performance on energy security over time under the dimensions of availability, affordability, efficiency, sustainability, and governance using 20 metrics and measured international performance across 18 countries from 1990 to 2010. Wang and Zhou (2017) evaluated energy security of 162 countries under a framework consisting of three energy security dimensions, namely, the Security of Energy Supply-Delivery dimension (SESD) with energy availability and infrastructure as two major factors, the Safety of Energy Utilization Dimension (SEUD) with equity of energy service and environmental sustainability of energy consumption as two major factors, and the Stability of the Political-Economic Environment Dimension (SPED) mainly involving a country's political and economic strength. The concept of "sustainable energy security" was proposed by Narula and Reddy (2016) suitable to assess energy security of developing countries where the energy system has been divided into "supply", "conversion and distribution", and "demand" sub-systems, each of which is further sub-divided into four dimensions of availability, affordability, efficiency and acceptability.

7.2   ENERGY SECURITY IN THE CONTEXT OF NEPAL

Nepal's energy resources are broadly divided into three categories: traditional, commercial and alternative sources. Traditional energy resources include biomass resources (fuelwood, agricultural residues, and animal wastes) used for energy production. All the energy resources with well-established market prices are grouped into commercial energy sources, such as petroleum products (petrol, diesel, aviation transportation fuel (ATF), kerosene, LPG), coal, and grid electricity, whereas

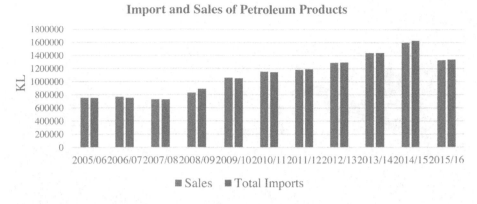

FIGURE 7.1 Import and sales of petroleum products of Nepal in2005/06–2015/16. (Source: NOC, 2017).

indigenous renewable energy resources fall under alternative sources. The high population growth and increasing economic growth of the country have increased the energy demand. With an increase in demand, the challenges of energy security and sustainability are becoming more pressing because the country relies heavily on imported energy. Nepal's energy use is primarily dominated by traditional energy sources, mainly biomass for domestic purposes which creates issues of deforestation, indoor air pollution and sustainability. Petroleum products such as petrol, diesel, kerosene and ATF are dominant in the transportation sector, while in industrial, residential and commercial sectors, petroleum products such as diesel, petrol, kerosene and LPG are used mostly for cooking, water boiling, and space heating purposes. In the transport sector, there are no alternatives for petroleum products, except electricity whose use in the transport sector in Nepal is insignificantly low. The demand for petroleum products is increasing at an alarming rate as seen by the imports and sales of petroleum products from figure 7.1. As a net importer of petroleum products, and from only one supplier country, India, Nepal becomes highly vulnerable to oil supply disruptions, oil price fluctuations and their effects. Also, the continued use of the petroleum products in the country highlights the issue of energy security and the equally important issue of sustainability.

The petroleum product price has very high volatility. The average annual prices of petroleum products, i.e., petrol, etc. in 2005/2006–2015/2016 as in Figure 7.2 shows an increasing trend.

Similarly, Nepal spends a huge amount of money on imports of petroleum products and electricity as the demand for energy is increasing. With the increase in price of petroleum products, the expenditure made on imports of petroleum products is becoming significantly higher with time, as seen in Figure 7.3. The expenditure made for petroleum products has been exceeding the value earned from exports in 2010/2011, which is not preferable for the economic health of the country.

Despite a huge potential for renewable energies such as hydropower, solar power and wind energy, the country is facing an acute power shortage of electricity. The imbalance between electricity demand and its supply from indigenous sources has

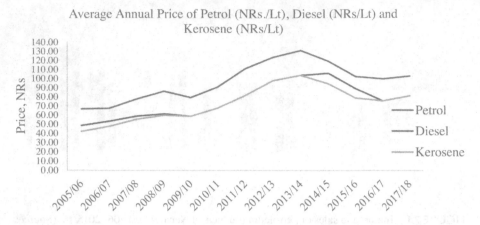

FIGURE 7.2 Average annual price of petrol, diesel, kerosene and LPG in 2005/2006–2017/2018 (Source: NOC, 2017)

made the country more dependent on import of electricity from India. As seen in Table 7.2, the import of electricity has increased drastically from 2005/2006 to 2016/2017 and has reached almost 35% of the total available energy in 2016/2017. Nepal has spent about 16 billion NRs. on import of electricity from India in 2016/2017 at an average power purchase rate of 7.36 NRs/kWh (NEA, 2017). The supply-demand gap of electricity is prevalent despite the increase in import of electricity hindering the economic development of the country.

Nepal was ranked 74th out of 94 countries in 2012, and 111th out of 129 countries in 2016 in terms of Energy Trilemma Index constructed by the World Energy Council that consists of energy security, energy equity and environmental sustainability (WEC, 2018). In a study of economic losses in the last decade due to load-shedding in Nepal by Timilsina et al. (2018), it was found that Nepal could not meet, on average, 20% of its total electricity demand in each year during the 2008–2016

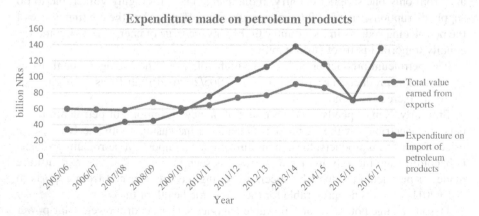

FIGURE 7.3 Expenditure made for petroleum products imports and total value earned from exports from 2005 to 2017. (Source: NOC, 2017; MoF, 2017).

TABLE 7.2
Indigenous Production and Imports of Electricity in Nepal in 2005/2006–2016/2017

Year	Indigenous Production		Imports, GWh		Available Energy, GWh
	GWh	% of Total Available Energy	GWh	% of Total Available Energy	
2005/2006	2514.69	90.43	266.23	9.57	2780.92
2006/2007	2722.99	89.23	328.83	10.77	3051.82
2007/2008	2760.73	86.65	425.22	13.35	3185.95
2008/2009	2774.33	88.61	356.46	11.39	3130.79
2009/2010	2713.09	80.94	638.68	19.06	3351.77
2010/2011	3164.32	82.01	694.05	17.99	3858.37
2011/2012	3432.56	82.15	746.07	17.85	4178.63
2012/2013	3467.94	81.44	790.14	18.56	4258.08
2013/2014	3368.35	71.86	1318.7	28.14	4687.1
2014/2015	3637.05	72.64	1369.8	27.36	5006.94
2015/2016	3299.46	64.99	1777.6	35.01	5077.14
2016/2017	4082.69	65.24	2175.0	34.76	6257.73

Source: NEA, 2017.

period which would have reduced Nepal's GDP by more than 6% during that period and estimated that Nepal has lost a total of US$11 billion value of GDP in 9 years during the 2008–2016 period due to electricity load shedding. Electricity generated from hydropower is the most promising energy sources of Nepal which can make the country self-sufficient in terms of fulfilling its energy demand along with reducing greenhouse gas (GHG) emissions. The government of Nepal declared 2016–2026 as "National Energy Crisis Prevention and Electricity Development Decade" in 2015 with the target of 10,000 MW by 2025, and it prepared an action plan consisting of activities covering (i) legal reform provisions, (ii) policy decisions, (iii) administrative decisions and procedural reforms, and (iv) structural provisions and reforms (ADB, 2017). Similarly, the sustainable development goals (SDGs) of Nepal aspire to give access to affordable, reliable, sustainable and modern energy for all by 2030, which is studied to be achieved by increasing hydroelectricity production at least 10-fold during the next 15 years (NPC, 2015). The SDG7 of Nepal is formulated under the following objectives of the Sustainable Energy for All (SE4ALL) initiative:(i) ensure universal access to modern energy services, (ii) double the share of renewable energy in the global energy mix, and (iii) double the rate of improvement in energy efficiency by 2030 (NPC, 2015). The Regulatory Indicators for Sustainable Energy (RISE) report shows that the energy access scores are higher for countries with renewable energy sources such as Nepal, Mali and Tanzania, and the countries that have lagged behind in renewable energy have a less diversified fuel mix (World Bank, 2014).

7.3   ENERGY SECURITY ASSESSMENT—A CASE STUDY OF NEPAL

Nepal has been facing many challenges in the development of the energy sector. Some of the energy- related issues in Nepal based on Water Energy Commission Secretariat WECS (2010), WECS (2013a), WECS (2013b) and **ADB** (2017) can be listed as follows:

1. Low access to modern energy
2. Poor diversification of energy resources
3. High reliance on traditional fuels
4. Energy crisis/increasing gap between energy demand and supply
5. High dependence on imported energy
6. High dependence on import for fossil fuel supply and inadequate storage capacity
7. Inadequate power supply systems
8. Affordability of energy sources, electricity and petroleum products, in particular
9. Absence of energy conservation/efficiency
10. Lack of infrastructure, adequate transmission lines
11. Lack of energy reliability and energy security, etc.

7.3.1   RESEARCH METHODS ADOPTED

In order to analyze the energy security context in Nepal, a study has been conducted where the methodology adopted is general research. Mainly exhaustive literature review was done of research papers, books, journal papers, reports, national, regional and international reports, theses, etc. on energy security from the period 2007–2018. Similarly, different national reports, research papers, data archives, newspaper articles, etc. were studied regarding national energy situation, energy data and energy policies. Energy security indicators were formulated using the historical data collected from different sources from 2005/2006 to 2016/2017, and from the data generated from scenario analysis from 2017/2018 to 2029/2030. The indicators were then normalized, weighted and aggregated to form a single energy security index. For this study, historical assessment of energy security was carried out from 2005/2006 to 2015/2016 in order to analyze the trend of different energy security indicators. The baseline year for energy modeling is taken as 2016/2017 and projection of energy supply and demand scenario is carried out from 2016/2017 to 2029/2030 in order to track the variation of energy security indicators in the future. The timeframe for the study is selected to match the SDG implementation and achievement timeframe of Nepal along with the targets of the National Energy Crisis and Electricity Development Decade 2016–2026 action plan. The free statistical software RStudio version 1.1.463 is used for statistical analysis. RStudio is available in open source and commercial editions (RStudio Team, 2015).

The energy model for the energy security assessment of Nepal from 2016/2017 to 2029/2030 for evaluation of energy security indicators and energy security index in the future under different scenarios was developed in the LEAP (Long-range Energy

Alternatives Planning) software. The LEAP modeling framework is selected for this study due to its compatibility with published energy demand data. In the Key Assumptions section of the LEAP Model, the following parameters are used on the basis of data compilation obtained from the Ministry of Finance (MoF, 2017) and the Central Bureau of Statistics (CBS, 2012):

   i. Population: 28.17 million: (Mountain: 6.7%, Hill: 43.0%, Terai: 50.3%, Rural: 55.9%, and Urban: 44.1%)
  ii. Average Household Size: 4.5
 iii. Number of Households given by Total Population divided by Average Household Size
  iv. GDP: 747.107 billion NPR
   v. Share of Agriculture GDP: 34.6%
  vi. Share of Commercial GDP: 55.06%
 vii. Share of Industrial GDP: 14.72%

The input energy consumption data for the base year 2016/2017 was calibrated by using the primary energy data collected by WECS (2013c) and multiplying the sectoral energy consumption data with appropriate desegregation ratios. The energy data are arranged in a bottom-up approach, i.e., the final energy consumption is estimated from the sum of total of energy consumed in the end-use devices in different sectors. The sectoral energy demand projection for residential and land transport end use was carried out using the formula as used in Shakya and Shrestha (2011):

$$ESD_{i,t} = \left(POP_t \middle/ POP_0\right)^{\alpha_{1i}} * \left(GDP_t \middle/ GDP_0\right)^{\alpha_{2i}} * ESD_{i,0}$$

The sectoral energy demand projection for commercial, industrial and agricultural sectors was carried out using the formula as used in Shakya and Shrestha (2011):

$$ESD_{i,t} = \left(VA_t \middle/ VA_0\right)^{\alpha_{3i}} * ESD_{i,0}$$

Where
   $ESD_{i,t}$ = level of service demand type i in year t for a sector or sub-region
   POP_t = population in year t
   GDP_t = aggregate GDP in year t
   VA_t = value added in the relevant sector in year t
   α_{1i} = population elasticity of service demand for service type i
   α_{2i} = GDP elasticity of service demand for service type i
   α_{3i} = sectoral value addition elasticity of service demand for service type i

The elasticities required by the equations are adopted from Shrestha and Nakarmi (2015), which are determined using historical data of energy consumption along with demographic and economic factors, as shown in Table 7.3.

TABLE 7.3
Elasticities for Sectoral Demand Projection

General	Population	GDP	GDP/Capita	VA (Industrial)	VA (Agricultural)	VA (Commercial)
Elasticity	1.053	0.1315	0.776	2.472	1.372	1.807

Source: (Shrestha and Nakarmi, 2015)

The energy demand of the transport sector in terms of passenger-kilometer for passenger transport and ton-kilometer for freight vehicles is calculated using the equation below based on Malla (2014):

$$TD_i = \sum_j \sum_k N_{j,k} * OF_{j,k} * VKT_{j,k} * LF_{i,j,k}$$

where N = number of registered vehicles
   OF = operational factor
   VKT = vehicle kilometer traveled (in km)
   LF = load factor (occupancy in passenger or ton)
   i, j, k = transport type, transport mode and energy type, respectively

Operation factor, load factor, and vehicle kilometer traveled per year are obtained from Malla (2014) to calculate passenger-kilometer and ton-kilometer.

On the supply side, the imports and sales of petroleum products for the year 2016/2017 are taken from the annual reports of Nepal Oil Corporation (NOC, 2017), and the data for electricity generation are based on annual reports of Nepal Electricity Authority (NEA, 2017). The transformation module consists of electricity generation, transmission and distribution losses, and own use module for the Business as Usual (BAU) Scenario with Petroleum Pipeline Imports module added for the Energy Security (ES) Scenario based on the Raxaul-Amlekhgunj Petroleum Pipeline project (NOC, 2016). The addition of hydropower plants is based on the plans and targets of different hydropower projects expected to come into operation within the end year 2030.

Two different scenarios were created, namely the BAU Scenario and the Energy Security ES Scenario. The BAU Scenario considers population growth of 1.35% and GDP growth rate of 4.2% taken as an average of economic growth rate of Nepal from 2007/2008 to 2016/2017 and no technology and/or policy interventions. The Scenario development is based on the Sustainable Development Goals of Nepal, Energy Crisis Reduction and Electricity Development Decade 2016-2026, Nepal's Energy Sector Vision 2050, National Energy Strategy of Nepal, 2013, Energy Efficiency Strategy of Nepal, 2075, etc. The ES Scenario considers population growth of 1.35% and GDP growth rate of 7% along with technology and/or policy interventions such as electrification, energy efficient technologies in residential, commercial, industrial, transport and agriculture sectors, shift toward electric mass transport system, and introduction of new energy sources for electricity generation and increase of renewable energy sources. The main assumptions of the scenarios are discussed below:

1. Business as Usual (BAU) Scenario: The main features of the Business as Usual Scenario are discussed below:
   - Population Growth Rate: 1.35% (CBS, 2012)
   - GDP Growth Rate: 4.2% (MoF, 2017)
   - Electricity Transmission and Distribution losses decrease from 22.9% in the base year 2016/2017 to 16% in the year 2030
   - Electricity generation capacity increased
2. Energy Security (ES) Scenario: The ES Scenario considers technology and policy intervention. Its main features are discussed below:
   A. Access to Electricity (NPC, 2017)
      - 100% electrification for lighting in the residential sector (80% electricity and 20% solar photovoltaic (PV) by 2030)
      - Phase-out of incandescent light and 100% LED penetration by 2030
      - Complete heating and water boiling in electricity in both rural and urban households by 2030
      - Penetration of efficient cooling, water pumping and electrical appliances (EnergyStar label) in both rural and urban households by 2030
   B. Access to Clean Cooking Stoves ICS, LPG, Electricity (NPC, 2017)
      - Traditional fuelwood stoves in rural areas will be completely replaced by ICS by 2030
      - Fuel mix for rural cooking as 70% ICS, 15% electricity, 10% biogas and 5% LPG in rural areas
      - Complete cooking in electricity in urban households by 2030
   C. Electrification of Industrial, Commercial and Agricultural Sectors
      - Electrification in power tiller, thresher and water pumping in the agriculture sector in the share of 33% by 2020, 66% by 2025 and 100% by 2030
      - Complete electrification in all end-uses of the commercial sector by 2030
      - Complete electrification for power motives, process heating, process cooling, boiling, lighting in the industrial sector following the trend 33% by 2020, 66% by 2025 and 100% by 2030 (Shakya, 2014)
   D. Electric Mass Transport and Electrification in the Transport Sector (NPC, 2017)
      - Penetration of electric buses in the transport sector with 10% by 2020, 15% by 2025 and 20% by 2030
      - Introduction of electric cars, 3-wheelers and 2-wheelers with 5% share each by 2030
   E. 10% of the total consumption of petrol and diesel in Nepal will have been replaced by biodiesel and bioethanol as mentioned in Biomass Energy Strategy of Nepal, 2017 (MoPE, 2017)
   F. Demand Side Management and Improved Energy Efficiency
      - Penetration of efficient technology in all sectors

G. Diversified Supply Portfolio
- 10, 000 MW by 2025 through an appropriate combination of storage, run-of-the-river, peaking run-of-the-river, and pump storage plants. The share of the plants being 40–50% from reservoir-based and pump storage, 15–20% peaking run-of-the-river, 23–30% run-of-the-river and 5–10% other energy security sources to end load-shedding from the power generated within the country in the long term (ADB, 2017). Storage hydro plants are endogenously added to maintain planning reserve margin of 25%
- 10 MW of electricity through the solid waste management from bio-gasifiers mentioned in Biomass Energy Strategy of Nepal, 2017 (MoPE, 2017)
- 25 MW of electricity generation by 2025 from sugarcane bagasse from sugarcane factories (GIZ/NEEP, 2016)
- 500 MW wind and 2000 MW solar addition for electricity generation by 2025 (NPC, 2018). According to the Solar and Wind Energy Resource Assessment (SWERA) project 2008 report, prepared by Alternative Energy Promotion Center (AEPC), the commercial potential of solar power for grid connection is 2100 MW and the potential of wind energy is 3000 MW in Nepal (AEPC, 2008)

H. Imports of Petroleum Products through Pipelines
- Diesel and gasoline imports from pipelines from Raxaul, India to Amlekhgunj, Nepal with a capacity of 2 million toe per year, starting from 2020. The import through pipeline is believed to ensure smooth supply of petroleum products, reduce any losses in transportation, prevent quality tampering and reduce transportation costs (NOC, 2017)

I. Electricity Transmission and Distribution Losses Reduction
- Electricity transmission and distribution losses reduce to 10% by 2030

In this study, energy security indicators, both simple and complex, found in the literature todate were listed and analyzed one-by-one in order to be suitable to use in the context of Nepal and selected based on data availability. Energy Indicators for Sustainable Development (EISD) developed by UNDP (IAEA, 2005), and the report of Water and Energy Commission Secretariat (WECS, 2013b) were also usedas reference. The indicators selected in this study for the assessment of energy security of Nepal are as below:

1. **Availability Dimension**
   - Diversity of Primary Energy Sources, Shannon-Weiner Index, AVL1
   - Net Energy Import Dependency, AVL2
   - Net Oil Import Dependency, AVL3
   - Number of Days of Stock of Petroleum Products, AVL4
   - Number of Days of Stock of LPG, AVL5
   - Self-Sufficiency or Share of Domestic Production of Primary Energy, AVL6
   - Annual Electricity Deficit, AVL7
   - Electricity Consumption per Capita, AVL8

2. **Affordability Dimension**
   - Value of Oil Imports per Unit of GDP, AFF1
3. **Accessibility Dimension**
   - Access to Electricity, ACS1
   - Access to Modern Cooking Fuel, ACS2
4. **Efficiency Dimension**
   - Total Final Energy Consumption per Capita, EFF1
   - Final Energy Intensity, EFF2
   - Oil Consumption per Capita, EFF3
   - Oil Consumption per Unit of GDP, EFF4
   - Residential Energy Intensity, EFF5
   - Industrial Energy Intensity, EFF6
   - Commercial Energy Intensity, EFF7
   - Agricultural Energy Intensity, EFF8
   - Transport Energy Intensity, EFF9
   - Electricity Transmission and Distribution Losses, EFF10
5. **Acceptability Dimension**
   - Non-carbon Fuel Portfolio, ACP1
   - Annual CO_2 Emissions per Unit of Energy Use, ACP2

After the selection of energy security indicators suitable in the context of Nepal, the indicators were normalized, weighted and aggregated to form a single energy security index that could be used to track changes in the country's performance over time. Normalization such as min–max, distance to reference, and standardization is done to transform the selected indicators having different units or on different scales into indicators with no or common unit/scale, so that they can be justifiably aggregated to form a composite index. The weights of the indicators can be assigned based on expert opinions collected from surveys, interviews, the Delphi method, the analytical hierarchy process, or other procedures such as equal weighting, principal component analysis, fuel or import share, etc. Aggregation involves combining the weighted indicators to form a single composite index. For this study, the normalization, weighting and aggregation methodology has been adopted from Martchamadol and Kumar (2013) where an "Aggregated Energy Security Performance Indicator" suitable for analysis of energy security at the national and provincial level with time series data of indicators has been developed and used in finding AESPI of Thailand. AESPI was developed by using 25 indicators of Energy Indicator for Sustainable Development (Martchamadol and Kumar, 2013). After identifying the group of indicators and determining weighting factors using the Principal Component Analysis (PCA) method, normalization of each relative indicator in a range of "0" representing the worst performance, to "10" representing the best performance is carried out. PCA method groups together individual indicators which are collinear to form a composite indicator that captures as much as possible of the information common to individual indicators such that the highest possible variation in the indicator set is explained through smallest possible number of factors (OECD 2008). For normalization, the indicators are separated into positive indicators whose high value represents high improvement in energy security such as Diversification Index, SWI, share of domestic production in TPES, Non-Carbon Fuel Portfolio, etc., and negative indicators whose high value of indicator implies low energy security such as Net Energy Import Dependency, Net Oil Import Dependency, Value of Oil Imports per unit of GDP, Annual Electricity Deficit %, Yearly Electricity Transmission and Distribution Losses %, Energy Consumption per Capita, Energy Intensity, and Oil Consumption per GDP. Positive indicators are normalized by using a scaling technique as given by the formula (Martchamadol and Kumar, 2013):

$$Max_i = Max\left(X_{ij}\ldots\ldots\ldots\ldots\ldots X_{in}\right);$$

$$\varnothing_{ij} = \frac{10 \times X_{ij}}{Max_i}$$

Where $\varnothing_{ij}$ is relative indicator i of year j,

X_{ij} is the value of positive indicator i of year j, and
Max_i is the maximum value of indicators i.

For normalization of the negative indicators, the inverse of the indicators (Y_{ij}) were taken in the above formula (Martchamadol and Kumar, 2013).

$$Y_{ij} = \frac{1}{X_{ij}}$$

The overall outcome of each group is given by its Group Index (GI) obtained as the rootmean square of all the relative indicators within group "k" of year "j" using the following formula (Gnansounou, 2008) :

$$GI_{kj} = \left(\sum \left(\varphi_{ij}\right)^2 / m\right)^{1/2}$$

Where, GI_{kj} is the Group Indicator k of year j,

φ_{ij} is the relative indicator I of year j, and
m is the number of indicators in each group.

AESPI is developed from GI_{kj} and group weighting factors, w_k, using the AESPI equation (Martchamadol and Kumar, 2013) :

$$AESPI_j = \frac{\sum \left(w_k \times GI_{kj}\right)}{\sum w_k}$$

Where $AESPI_j$ is Aggregated Energy Security Performance Indicator of year j;

GI_{kj} is Group Indicator k of year j, and
M is the number of group indicators.

7.3.2 RESULTS AND DISCUSSIONS

7.3.2.1 Scenario Analysis

7.3.2.1.1 Final Energy Demand Projection

The final energy demand projection in the Business As Usual (BAU) Scenario shows an increasing trend with the base year 2017 and the end year 2030 values being 502.79 PJ and 824.79 PJ, respectively. The energy consumption is seen to be the highest in the residential sector and the lowest in the agricultural sector throughout the study period in the BAU Scenario, while the transport sector demand is seen to be rising in an alarming rate with the end year value nearly four times the base year value. The end year demand is nearly two-fold times greater than the base year value. The final energy demand in the BAU Scenario projected through 2016/2017–2029/2030 is shown in Table 7.4.

The final energy demand projection in the Energy Security (ES) Scenario shows an increasing trend with the base year 2017 and the end year 2030 values being 502.79 PJ and 519.47 PJ, respectively. The energy consumption is seen to be the highest in the residential sector and the lowest in the agricultural sector in the ES Scenario as well, while the transport sector demand is seen to be rising at an alarming rate with the end year value nearly 3.5 times the base year value. The residential energy demand is decreasing overthe years along with the commercial energy demand. The final energy demand in the ES Scenario projected through 2016/2017–202920/30 is shown in Table 7.5.

The complete electrification in the residential and commercial sector along with penetration of efficient technologies for lighting, cooking, heating, cooling, electric appliances, etc. has decreased the final energy consumption in the ES Scenario. A comparative illustration of the final energy demand projection of BAU and ES Scenarios is shown in Figure 7.4.

TABLE 7.4
Final Energy Demand in BAU Scenario

Branches, PJ	2017	2020	2023	2026	2030	2030/2017
Residential	385.85	401.69	418.17	435.34	459.33	1.19
Commercial	18.17	22.26	27.26	33.40	43.78	2.41
Transport	51.80	71.38	98.40	135.71	208.46	4.02
Agricultural	5.36	6.57	8.04	9.85	12.92	2.41
Industrial	41.62	50.99	62.46	76.52	100.30	2.41
Total	502.79	552.87	614.34	690.82	824.79	1.64

TABLE 7.5

Final Energy Demand in ES Scenario

Branches, PJ	2017	2020	2023	2026	2030	2030/2017
Residential	385.85	356.06	324.80	290.41	239.39	0.62
Commercial	18.17	18.20	18.66	18.41	16.31	0.90
Transport	51.80	69.23	93.23	126.11	190.63	3.68
Agricultural	5.36	5.94	6.85	7.87	9.37	1.75
Industrial	41.62	44.94	50.32	55.96	63.77	1.53
Total	502.79	494.37	493.87	498.76	519.47	1.03

7.3.2.1.2 Fuel Mix Projection

The fuel mix in the final energy demand projection of the BAU Scenario through 2016/2017–2029/2030 is shown in Table 7.6.

The share of fuel wood is the highest among all fuels followed by the share of diesel from commercial energy sources in the BAU Scenario. The fuel mix in the final energy demand projection in the ES Scenario through 2016/2017–2029/2030 is shown in Table 7.7.

In the ES Scenario, the highest share of fuel is that of wood in both the base year 2017 and end year 2030; however, the share of fuel wood is decreasing from the base year 2017 to end year 2030 in the ES Scenario. The share of diesel is increasing from the base year to the end year along with the rapid increment in the share of electricity in the study period. As the ES scenario creates a completely electrified residential and commercial sector, the share of electricity increases, and the replacement of fuel wood by electricity for cooking, heating, water boiling, etc. decreases the consumption of fuel wood. The share of LPG is declining in the ES scenario, while kerosene, coal and furnace oil are completely phased out in the end year in the ES scenario. Since these fuels are imported, the reduction in the primary energy demand of these

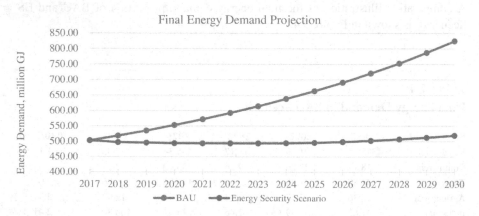

FIGURE 7.4   Final energy demand projection of BAU vs. ES scenarios. (Source: (Darlamee 2019)

TABLE 7.6
Fuel Mix in the Primary Energy Demand for BAU Scenario

Fuels, PJ	2017	2020	2023	2026	2030
Electricity	17.27	19.64	22.49	25.91	31.60
Gasoline	13.19	18.33	25.52	35.60	55.57
Jet Kerosene	5.94	8.34	11.71	16.44	25.84
Kerosene	0.78	0.84	0.90	0.97	1.08
Diesel	46.24	61.27	81.38	108.34	159.25
Liquefied Petroleum Gas (LPG)	15.34	16.72	18.33	20.23	23.30
Oil	0.01	0.01	0.01	0.01	0.02
Coal Bituminous	18.75	22.97	28.13	34.47	45.18
Wood	337.12	354.62	373.67	394.52	425.63
Biogas	5.85	6.09	6.34	6.60	6.96
Ethanol	-	-	-	-	-
Animal Wastes	24.52	25.53	26.57	27.66	29.19
Solar	0.17	0.17	0.18	0.19	0.20
Biomass	17.20	17.91	18.65	19.41	20.48
Micro Hydro	0.41	0.43	0.45	0.47	0.49
Total	502.79	552.87	614.34	690.82	824.79

TABLE 7.7
Fuel Mix in the Primary Energy Demand for ES Scenario

Fuels, PJ	2017	2020	2023	2026	2030
Electricity	17.27	35.66	61.93	95.05	153.31
Gasoline	13.19	15.84	18.78	21.79	25.31
Jet Kerosene	5.94	8.34	11.71	16.44	25.84
Kerosene	0.78	0.65	0.49	0.30	-
Diesel	46.24	57.14	72.54	91.68	122.84
Liquefied Petroleum Gas (LPG)	15.34	13.29	10.91	7.95	2.75
Oil	0.01	0.01	0.01	0.00	-
Coal Bituminous	18.75	17.02	14.13	9.45	-
Wood	337.12	303.62	266.28	223.63	156.83
Biogas	5.85	5.58	5.27	4.94	4.43
Ethanol	-	-	-	1.62	10.98
Animal Wastes	24.52	19.99	15.04	9.65	1.74
Solar	0.17	2.65	5.33	8.24	12.46
Biomass	17.20	14.27	11.13	7.70	2.66
MicroHydro	0.41	0.33	0.32	0.32	0.31
Total	502.79	494.37	493.87	498.76	519.47

TABLE 7.8

CO_2 Emissions in the BAU and ES Scenarios

Scenarios	2017	2020	2023	2026	2030
Energy Security (ES)	9.42	9.92	10.68	11.58	12.93
Business As Usual (BAU)	9.42	11.61	14.50	18.31	25.41
Ratio	0	0.85	0.74	0.63	0.51

fuels is beneficial for the country, as it reduces import dependency. However, the share of diesel and gasoline is increasing in the ES Scenario as well since the transport sector relies on these imported fuels.

7.3.2.1.3 Emissions Projection

The global warming potential (GWP) is a measure of how much energy the emissions of 1 ton of a gas will absorb over a given period of time, relative to the emissions of 1 ton of carbon dioxide (CO_2). The larger the GWP, the more that a given gas warms the earth compared to CO_2 over that time period (EPA, 2019). The emissions of CO_2(100-year GWP) for 2016/2017–2029/2030 as given by the LEAP energy model in the BAU and ES Scenarios are shown in Table 7.8.

The 100-year GWP (global warming potential calculated for a 100-year time-frame) is estimated to be 9.42 million Metric tons of CO_2 equivalents in the base year 2016/2017. The emission results from the BAU and ES Scenarios show an increasing emission trend in both scenarios as seen in Figure 7.5, reaching 25.41 million Metric tons of CO_2 equivalents and 12.93 million Metric tons of CO_2 equivalents in the end year 2030 in the BAU and ES Scenarios, respectively.

7.3.2.1.4 Electricity Demand Projection

Electricity from the indigenous hydro resources can strengthen the energy security condition of Nepal by reducing dependency. The electricity demand projection for 2016/2017–2029/2030 in the BAU and ES Scenarios is shown in Table 7.9.

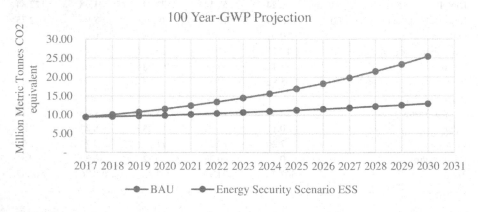

FIGURE 7.5 CO_2 emissions in BAU and ES scenarios. (Source: (Darlamee 2019)

TABLE 7.9

Electricity Demand Projection in BAU and ES Scenarios

Scenarios, GWh	2017	2020	2023	2026	2028	2030
Energy Security (ES)	4,796.53	9,906.39	17,203.80	26,402.26	33,812.73	42,585.89
Business As Usual (BAU)	4,796.53	5,456.23	6,247.22	7,198.35	7,938.55	8,777.83

The result shows that the electricity demand is increasing in both scenarios, but the electricity demand is increasing rapidly in the ES Scenario as seen in Figure 7.6 which is justifiable by the fact that it features rapid electrification in almost all the service end-use demand of all the consumption sectors. The demand for electricity in the ES Scenario for the end year 2030 is nearly 10-fold the electricity demand in the base year. The complete electrification by 2030 target requires the electricity demand to be 42,585.89 GWh in the ES Scenario in the end year 2030, which is about five times greater than the end year electricity demand of 8,777.83 GWh in the BAU Scenario.

For the fulfillment of high electricity demand in the scenario years 2017–2030 in the ES Scenario, electricity generation has to be increased or imports of electricity have to be increased. The fuel used for electricity generation produced domestically in the ES Scenario is shown in Table 7.10.

7.3.2.1.5 Petroleum Products Demand Projection

The petroleum products demand projection for 2016/2017–2029/2030 in the BAU Scenario is shown in Table 7.11.

The results show that the demand for petroleum products increases with every year in the BAU Scenario, with the highest share of diesel followed by LPG. The demand

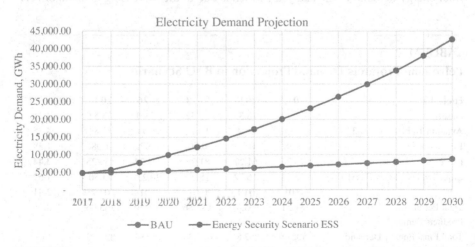

FIGURE 7.6 Electricity demand projection in BAU and ES scenarios. (Source: (Darlamee 2019)

TABLE 7.10

Diversity of Electricity Generation (Indigenous Production) in ES Scenario

Branches, GWh	2017	2020	2023	2026	2028	2030
Base Load Hydro	3,971.21	10,104.59	10,902.67	16,484.02	20,932.7	24,383.8
Peak Storage Hydro	111.20	1,008.19	476.98	-	83.19	3,194.18
Thermal	0.28	-	-	-	-	-
Solar	-	724.61	6,883.58	9,615.68	12,210.7	14,223.9
MSW Gasifier Plants	-	-	-	39.25	49.84	58.06
Wind	-	665.46	632.17	2,207.68	2,803.49	3,265.70
Bagasse Plant	-	-	-	-	0.31	11.90
Micro Hydro	-	-	2,099.49	2,848.54	3,045.01	3,145.66
Total	4,082.69	12,502.86	20,994.88	31,195.17	39,125.3	48,283.3

for jet kerosene or Aviation Turbine Fuel (ATF) in the end year 2030 increases by 4.35 times the base year value. The demand for petroleum products in the end year covers 32.14% of the total final energy demand in the end year 2030. The petroleum products demand projection for 2016/2017–2029/2030 in the ES Scenario is shown in Table 7.12.

The results show that the demand for petroleum products increases with year in the ES Scenario as well, with the highest share of diesel followed by ATF. The demand for total petroleum products increases by 2.17 times the base year demand, while the demand for ATF in the end year 2030 increases by 4.35 times the base year value. While the demand for LPG decreases along with the demand for kerosene and furnace oil throughout the years, the demand for petroleum products in the end year 2030 is more than twice the demand at the base year and covers 34.02% of the total final energy demand in the end year, which is due to the increasing demand of ATF

TABLE 7.11

Petroleum Products Demand Projection in BAU Scenario

Fuels, PJ	2017	2020	2023	2026	2030	2030/2017
Gasoline	13.19	18.33	25.52	35.60	55.57	4.21
Aviation Turbine Fuel (ATF)	5.94	8.34	11.71	16.44	25.84	4.35
Kerosene	0.78	0.84	0.90	0.97	1.08	1.38
Diesel	46.24	61.27	81.38	108.34	159.25	3.44
Liquefied Petroleum Gas (LPG)	15.34	16.72	18.33	20.23	23.30	1.52
Oil	0.01	0.01	0.01	0.01	0.02	2.41
Total Petroleum Products Demand	81.51	105.51	137.86	181.59	265.06	3.25
Total Final Energy Demand	502.79	552.87	614.34	690.82	824.79	
% of Final Energy Demand	16.21	19.08	22.44	26.29	32.14	

TABLE 7.12

Petroleum Products Demand Projection in ES Scenario

Fuels, PJ	2017	2020	2023	2026	2030	2030/2017
Gasoline	13.19	15.84	18.78	21.79	25.31	1.92
Aviation Turbine Fuel (ATF)	5.94	8.34	11.71	16.44	25.84	4.35
Kerosene	0.78	0.65	0.49	0.30	-	-
Diesel	46.24	57.14	72.54	91.68	122.84	2.66
Liquefied Petroleum Gas (LPG)	15.34	13.29	10.91	7.95	2.75	0.18
Oil	0.01	0.01	0.01	0.00	-	-
Total Petroleum Products Demand	81.51	95.26	114.44	138.17	176.74	2.17
Total Final Energy Demand	502.79	494.37	493.87	498.76	519.47	
% of Total Energy Demand	16.21	19.27	23.17	27.70	34.02	

and diesel, which are the major drivers in the transportation sector. As the transport sector demand is rising at a high rate, the demand for diesel and ATF increases accordingly. The replacement of diesel buses by electric buses by 20% in 2030 is not sufficient to decrease the share of diesel in the final energy demand.

7.3.2.2 Energy Security Indicators Formulation

The indicators were calculated from the data of the metrics collected from different national and international reports for 2005/2006–2015/2016. The value of the indicators in 2016/2017–2029/2030 is calculated by using the value of the metrics extracted from the scenario analysis. The formulas used to calculate the value of the indicators are listed in the Appendix at the end of the chapter. In the "Availability Dimension," it can be seen that out of eight indicators, in the ES scenario, capacity addition of hydro and alternatives accompanied by the introduction of modern biomass fuels such as municipal waste gasification to electricity generation, sugarcane bagasse cogeneration plants and biodiesel strengthens the Energy Security condition of Nepal by increasing the value of these indicators: Diversity Index of Primary Energy Sources, AVL1; Share of Domestic Production of Primary Energy, AVL6; and Electricity Consumption per Capita, AVL8, and by decreasing the value of these indicators: Annual Electricity Deficit, AVL7; Number of Days of Stock of Petroleum Products, AVL4; and Number of Days of Stock of LPG, AVL5. The increase in strategic storage capacity of oil reduces the vulnerabilities such as sudden supply disruptions and price hikes. However, since the transportation sector is dominated by petroleum, which is totally imported from other countries, an increase in Net Energy Import Dependency, AVL2, and Net Oil Import Dependency, AVL3 can be seen, which still poses risks.

In the "Affordability Dimension," the Value of Energy Imports per Unit of GDP of Nepal, AFF1, is increasing and the highest is in 2013/2014 with almost 18% of GDP spent on imports of petroleum products. The decrease in 2015/2016 is due to the decrease in the import of petroleum products caused by the informal blockade imposed by India. As both the import volume and cost of petroleum products are increasing in the scenario analysis period, it is seen that the value of imports of petroleum products per unit of GDP is also increasing and it reaches the highest value of 0.18 in the end year in the ES Scenario implying that 18.3% of GDP of Nepal at the end year will be spent on import of petroleum products only.

In the "Accessibility Dimension," in the BAU and ES scenarios the Access to Electricity, ACS1, and Access to Modern Cooking Fuel, ACS2, follow an increasing trend with increase in access to national grid electricity along with electrification from alternative sources such as solar and mini/micro hydro. The BAU Scenario follows the present trend in terms of households using clean fuel for cooking, while in the ES Scenario, the traditional stoves are replaced by cleaner ICS, LPG, biogas and electricity.

In the "Acceptability Dimension," the addition of hydropower plants, solar, and wind for electricity generation in the ES Scenario strengthens the energy security condition of Nepal by improving the Non-Carbon Fuel Portfolio, ACP1. The CO_2 Emissions Generated per Unit of Energy Use, ACP2, shows an increasing trend in both BAU and ES Scenario, but the growth is less in ES Scenario compared to that in the BAU Scenario.

In the "Efficiency Dimension," the penetration of efficient technologies and appliances in the residential, commercial, agricultural sectors, and use of high fuel economy vehicles and electric vehicles in the transportation sector in the ES Scenario, the final energy consumption decreases which in turn decreases the value of indicators such as Total Final Energy Consumption per Capita, EFF1; Final Energy Intensity, EFF2; Residential Energy Intensity, EFF5; Industrial Energy Intensity, EFF6; Commercial Energy Intensity, EFF7; and Agricultural Energy Intensity, EFF8. The value of the indicator Oil Consumption per Capita, EFF3, and Oil Consumption per Unit of GDP, EFF4, shows an increasing trend in the BAU and ES Scenarios.

7.3.2.3 Principal Component Analysis

7.3.2.3.1 PCA Testing

The standardized indicators were tested for sampling adequacy and correlation testing by Kaiser– Meyer–Olkin (KMO) and Bartlett's tests. KMO is a measure of sampling adequacy that compares the magnitudes of the observed correlation coefficients to the magnitudes of the partial correlation coefficients (OECD 2008). The overall measure of sampling adequacy (MSA) obtained was 0.5, which fulfilled the criteria 0. 5 < KMO < 1 (Child, 1990) in the KMO test, and in the Bartlett's test the significance level of chi-square results were lower than 0. 05 (Dogbegah et al., 2011), indicating adequacy of the PCA method. After passing the KMO and Bartlett's tests of sphericity, the "correlation matrix (R)", "eigenvalues" and "factor loading"are obtained from PCA. The number of principal components with eigenvalue greater than one gives the number of groups of indicators (**OECD,** 2008). Only the first three principal components were found to have eigenvalues greater than one, which is shown in Table 7.13.

Hence, the first three principal components were only retained and the indicators were grouped accordingly under three groups for further analysis. The weighting factors of each group (w_k) are given by the rotation of percentage of variance divided

TABLE 7.13

Eigenvalues of the Principal Components from PCA

BAU Scenario		ES Scenario	
Principal Components	Eigenvalues	Principal Components	Eigenvalues
PC1	11.16008	PC1	14.1336
PC2	8.299274	PC2	4.884704
PC3	1.287118	PC3	1.433691

by the rotation of cumulative of variance obtained from "Varimax" rotation (Doukas et al., 2012). The weighting factors for the Group Indicator under the BAU Scenario are shown in Table 7.14.

The indicators with factor loading value close to 1 (or –1) are listed under the same group (Doukas et al., 2012). The rotated component matrix with three of the retained principal components in the BAU scenario is shown in Table 7.15.

The rotated component matrix shows that the indicators can be categorized into three groups with 13 indicators under group 1, 8 indicators under group 2, and 2 indicators under group 3. The rotated component matrix with 3 of the retained principal components in the ES Scenario is shown in Table 7.16.

Group Index (GI) is obtained as the root mean square of all the normalized indicators (Gnansounou, 2008). The GI formation and weighted value of GI in the BAU and the ES scenarios are shown in Table 7.17.

7.3.2.4 Energy Security Index

After normalizing all the energy security indicators on the scale of 0 to 10, first an equal weighting factor for each indicator was used and also the weight obtained from PCA was used to aggregate the indicators to form a composite energy security index of Nepal from 2005 to 2030. The calculated energy security index of Nepal for the period of 2005 to 2030 under the BAU and ES Scenarios is shown in Table 7.18.

TABLE 7.14

Weighting Factors of Groups Obtained from PCA in the BAU and ES Scenarios

	BAU Scenario			ES Scenario		
Component, i	Eigenvalues	% of Variance Explained, Var_i	Weighting Factor, w_i	Eigenvalues	% of Variance Explained, Var_k	Weighting Factor, w_k
1	14.31	0.46	0.51	14.31	0.44	0.49
2	4.68	0.36	0.40	4.68	0.36	0.41
3	1.43	0.08	0.09	1.43	0.09	0.10
	Cumulative % ($\sum Var_k$)		0.90	Cumulative % ($\sum Var_k$)		0.89

TABLE 7.15

Rotated Component Matrix Loadings and Group of Indicators for BAU Scenario

Indicators	PC1	PC2	PC3
EFF10 (Electricity Transmission and Distribution Losses)	−0.96	0.16	0.1
EFF2 (Final Energy Intensity)	−0.77	0.52	0.04
EFF6 (Industrial Energy Intensity)	0.69	−0.43	0.53
EFF7 (Commercial Energy Intensity)	0.72	−0.12	0.45
EFF8 (Agricultural Energy Intensity)	0.79	−0.08	0.34
EFF9 (Transport Energy Intensity)	0.69	0.15	0
AVL7 (Annual Electricity Deficit)	0.82	0.16	0.27
EFF3 (Oil Consumption per Capita)	0.94	−0.28	0.16
EFF4 (Oil Consumption per Unit GDP)	0.97	−0.12	0.15
AVL2 (Net Energy Import Dependency)	0.85	0.51	−0.05
AVL3 (Net Oil Import Dependency)	0.98	−0.06	0.18
AFF1 (Value of Oil Imports per Unit of GDP)	0.92	0.22	0.25
AVL6 (Share of Domestic Production of Primary Energy)	0.95	0.27	0.04
ACP2 (Annual CO_2 Emissions per Unit of Energy Use)	0.36	−0.88	0.23
EFF1 (Total Final Energy Consumption per Capita)	0.27	−0.94	−0.21
AVL4 (Number of Days of Stocks of Petroleum Products)	0.44	−0.82	0.29
AVL8 (Electricity Consumption per Capita)	−0.15	0.98	0
AVL1 (Diversity of Primary Energy Sources)	0.1	0.97	−0.13
ACP1 (Non-carbon Fuel Portfolio)	0.23	0.94	−0.06
ACS1 (Access to Electricity)	0.47	0.87	−0.1
ACS2 (Access to Modern Cooking Fuel)	0.3	0.95	−0.04
EFF5 (Residential Energy Intensity)	−0.43	−0.49	−0.56
AVL5 (Number of Days of Stock of LPG)	0.17	−0.5	0.74

As seen from the above table, the energy security index in the BAU Scenario follows a declining trend using both equal weighting and PCA weighting methods. However, it is seen that the decline in the value of energy index in the scenario analysis results in almost constant linear progression in the BAU Scenario. The energy security index decreases from 7.22 in 2005 to 5.96 in 2030 using equal weighting, and the value decreases from 5.72 in 2005 to 4.42 in 2030 using PCA weighting.

A sharp decline can be seen from Figure 7.7 in the years 2008/2009 and 2009/2010 because of the increase in electricity deficit in 2009/2010 and the increase in imports of petroleum products from the same year. The lowest values of energy security index are seen in the years 2015/2016 and 2018/2019, which is around 5.8 in the BAU Scenario. The country faces a huge electricity deficit, and import of petroleum products and electricity to meet the supply shortage in these years, along with increase in energy consumption and expenditure on imports of energy make the country vulnerable to risks.

The energy security index is seen to be the least at the base year 2016/2017 as seen in Figure 7.8, which can be attributed to the above causes and the energy security index starts increasing from 2016/2017 in the ES Scenario reaching the maximum

TABLE 7.16

Rotated Component Matrix Loadingsand Group of Indicators for ES Scenario

Indicators	PC1	PC2	PC3
EFF10 (Electricity Transmission and Distribution Losses)	0.95	−0.26	0.1
EFF2 (Final Energy Intensity)	0.95	−0.3	0.1
EFF5 (Residential Energy Intensity)	0.95	−0.22	0.14
AVL2 (Net Energy Import Dependency)	−0.78	0.15	0.11
AVL5 (Number of Days of Stock of LPG)	0.9	0.15	0.16
AVL1 (Diversity of Primary Energy Sources)	0.74	−0.62	−0.07
ACP1 (Non-carbon Fuel Portfolio)	0.95	−0.25	0.08
ACS1 (Access to Electricity)	0.68	−0.7	−0.03
ACS2 (Access to Modern Cooking Fuel)	0.9	−0.42	0.02
AVL8 (Electricity Consumption per Capita)	0.96	−0.26	0.08
ACP2 (Annual CO_2Emissions per Unit of Energy Use)	−0.52	0.8	0.2
EFF1 (Total Final Energy Consumption per Capita)	0.17	0.58	0.48
EFF6 (Industrial Energy Intensity)	0.05	0.96	0.09
EFF7 (Commercial Energy Intensity)	0.45	0.76	−0.23
AVL7 (Annual Electricity Deficit)	−0.29	0.65	0.03
EFF3 (Oil Consumption per Capita)	−0.61	0.76	0.19
EFF4 (Oil Consumption per Unit of GDP)	−0.42	0.83	0.31
AVL3 (Net Oil Import Dependency)	−0.65	0.73	0.15
AVL4 (Number of Days of Stock of Petroleum Products)	−0.5	0.79	0.13
AVL6 (Share of Domestic Production of Primary Energy)	−0.41	0.84	0.25
AFF1 (Value of Oil Imports per Unit of GDP)	−0.39	0.83	−0.1
EFF8 (Agricultural Energy Intensity)	0.47	0.45	0.63
EFF9 (Transport Energy Intensity)	0.05	0.04	0.96

value in the end year 2030. This shows that the scenario developed with policy measures such as electrification, penetration of efficient technologies, demand side management and energy efficiency strategies, diversified supply portfolio, increased share of hydro and renewables helps in improving the energy security of the country in an ideal situation where a sudden crisis due to natural disasters, wars, and alike does not occur. Nonetheless, the increase in imports of petroleum products due to the dependency on them, especially for the transportation sector, weighs down the energy security situation. Thus, as it is evident that the petroleum products will still dominate the energy security situation of the country, the country should focus more on oil supply security by diversifying the suppliers, diversifying the fuel mix in the transport sector, building resilience capacity against sudden price shocks, putting emphasis on strategic petroleum reserves, etc (**Bajracharya and Darlamee 2019**).

7.3.2.5 Environment Security

Basically, entire study and analysis of this study are focused on the energy security context. It is a universal truth that energy security and environmental security come

TABLE 7.17

Group Index (GI) Formation in the BAU and ES Scenarios

| | GIFormation in BAU Scenario | | | | | | GIFormation in ES Scenario | | | | | |
Year	GI1	w1*GI1	GI2	w2*GI2	GI3	w3*GI3	GI1	w1*GI1	GI2	w2*GI2	GI3	w3*GI3
2005/06	7.98	4.07	2.017	0.807	9.34	0.84	4.42	2.17	8.53	3.50	8.51	0.85
2006/07	8.70	4.44	2.107	0.843	8.44	0.76	4.42	2.17	9.25	3.79	8.54	0.85
2007/08	8.65	4.41	2.100	0.840	8.28	0.74	4.52	2.22	8.91	3.65	9.86	0.99
2009/10	7.36	3.75	1.937	0.775	7.10	0.64	4.54	2.22	8.30	3.40	7.43	0.74
2010/11	7.84	4.00	1.999	0.800	6.91	0.62	4.68	2.29	7.61	3.12	6.04	0.60
2011/12	6.59	3.36	1.833	0.733	7.74	0.70	4.79	2.35	7.50	3.08	9.23	0.92
2012/13	6.59	3.36	1.834	0.734	6.99	0.63	5.11	2.50	6.28	2.58	7.44	0.74
2013/14	6.71	3.42	1.850	0.740	6.89	0.62	5.12	2.51	6.19	2.54	7.49	0.75
2014/15	6.38	3.25	1.804	0.722	6.28	0.56	5.23	2.56	5.90	2.42	8.53	0.85
2015/16	6.68	3.41	1.846	0.739	6.45	0.58	5.20	2.55	5.55	2.27	7.92	0.79
2016/17	6.00	3.06	1.749	0.700	6.26	0.56	5.14	2.52	6.12	2.51	7.17	0.72
2017/18	6.01	3.06	1.751	0.700	6.24	0.56	5.28	2.59	4.90	2.01	7.02	0.70
2018/19	5.91	3.02	1.737	0.695	6.23	0.56	5.47	2.68	4.89	2.01	7.17	0.72
2019/20	5.91	3.01	1.736	0.694	7.24	0.65	5.62	2.75	4.91	2.01	7.32	0.73
2020/21	5.91	3.01	1.736	0.694	7.17	0.65	5.96	2.92	5.09	2.09	7.49	0.75
2021/22	5.91	3.01	1.736	0.695	7.11	0.64	6.55	3.21	5.02	2.06	7.71	0.77
2022/23	5.92	3.02	1.738	0.695	7.05	0.63	6.94	3.40	5.01	2.05	7.83	0.78
2023/24	5.93	3.03	1.740	0.696	6.98	0.63	7.24	3.55	5.04	2.06	7.96	0.80
2024/25	5.95	3.04	1.742	0.697	6.93	0.62	7.57	3.71	5.05	2.07	8.08	0.81
2025/26	5.97	3.05	1.745	0.698	6.87	0.62	7.92	3.88	5.07	2.08	8.20	0.82
2026/27	6.00	3.06	1.749	0.700	6.81	0.61	8.29	4.06	5.10	2.09	8.31	0.83
2027/28	6.03	3.07	1.754	0.701	6.76	0.61	8.70	4.26	5.15	2.11	8.44	0.84
2028/29	6.06	3.09	1.758	0.703	6.71	0.60	9.18	4.50	5.20	2.13	8.57	0.86
2029/30	6.10	3.11	1.764	0.706	6.66	0.60	9.89	4.85	5.28	2.16	8.71	0.87

together. Therefore, a brief thought on the environmental security of Nepal is also included.

The notion of environmental security is context-specific, but it generally examines threat posed by environmental events and trends (caused by nature or human process) to individuals, communities, or nations. They entail a wide range of issues such as resource access, equity, economics, land tenure, and property rights.

Environmental security is one of the key policy concerns in developing counties such as Nepal where the livelihood of most people is dependent on natural environmental resources such as forests. The Constitution of Nepal explicitly and authoritatively provisions that the right to a clean environment shall be a fundamental right, and the Constitution ensures that the victim of environment pollution "shall have the right to" seek compensation from polluters.

In developing countries such as Nepal, most of the population is directly or indirectly dependent on natural resources and ecosystem services such as food, firewood, fodder, and timber from forests. With the increased rate of environmental degradation and climate change, Nepal is witnessing decreasing environmental services

TABLE 7.18
Energy Security Index of Nepal under BAU and ES Scenarios (2005–2030)

Year	BAU Scenario		ES Scenario	
	Equal Weighting	PCA Weighting	Equal Weighting	PCA Weighting
2005/06	7.22	5.72	6.145	6.51
2006/07	7.75	6.04	6.698	6.81
2007/08	7.67	5.99	6.604	6.86
2008/09	7.14	5.53	6.112	6.37
2009/10	6.70	5.17	5.684	6.02
2010/11	6.97	5.42	5.913	6.35
2011/12	6.43	4.79	5.251	5.82
2012/13	6.42	4.73	5.220	5.80
2013/14	6.51	4.78	5.211	5.84
2014/15	6.24	4.54	4.977	5.62
2015/16	6.45	4.73	5.172	5.75
2016/17	5.88	4.32	4.649	5.30
2017/18	5.91	4.33	4.751	5.40
2018/19	5.90	4.27	4.858	5.50
2019/20	6.15	4.36	5.188	5.76
2021/22	6.09	4.35	5.470	6.04
2022/23	6.08	4.35	5.660	6.24
2023/24	6.07	4.35	5.823	6.41
2024/25	6.05	4.36	5.993	6.59
2025/26	6.04	4.36	6.169	6.78
2026/27	6.04	4.37	6.362	6.98
2027/28	6.03	4.38	6.579	7.21
2028/29	6.03	4.40	6.839	7.49
2029/30	5.96	4.42	7.189	7.88

such as water (for drinking, irrigation, or hydropower), firewood that increasingly causes social conflict within communities and, sometimes, with neighbors that affect the community or national security. In Nepal, natural resources and environmental series are important factors for employment and economic growth. Most people who are dependent on income and employment from the primary sector such as agriculture, forestry, and tourism would be affected by the change in environmental change. Nepal's national economy is mainly dependent on these sectors and may affect Nepal's economic prosperity. Besides, environmental degradation can expose millions of people to health threats and weaken the quality of human capital required for the development. Increasing the impact of climate change with extreme weather events has an unprecedented impact on the loss of people, physical assets, and environmental damages. All these may have serious implications on Nepal's economic prosperity and overall national security.

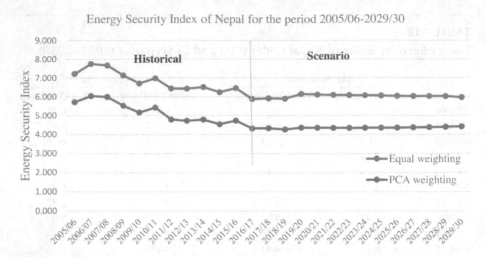

FIGURE 7.7 Energy security index of Nepal in BAU Scenario (2005–2030). (Source: (Darlamee 2019)

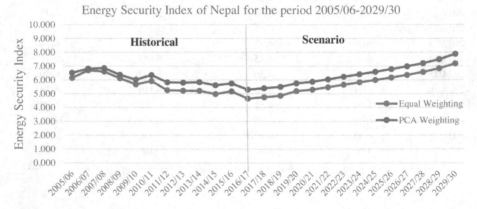

FIGURE 7.8 Energy security index of Nepal in ES Scenario (2005–2030). (Source: (Darlamee 2019)

7.4 CONCLUSION

The formulation of indicators using the historical data shows that apart from traditional energy sources, the diversity in fuel mix is very low, and it has not yet been improved in the period of 10 years from 2005/2006 to 2015/2016. Nepal is fully dependent on imported fossil fuel-based energy sources and electricity as well, which has made the country vulnerable to supply disruptions and price shocks. The high value of energy imports to GDP reflects high economic vulnerability to oil price shocks. It can be seen that the value of energy imports per unit of GDP of Nepal is increasing and the highest was in 2013/2014. Similarly, the stock days of petroleum products of Nepal that can act against short-term supply disruptions are declining and was just up to 11 days in

2015/2016, which is very low compared to the standard 90 days. On the electricity sector, Nepal faced electricity deficit throughout the period of 2005/2006–2016/17, while the imports of electricity increased in the 2014/2015 and 2015/2016 which increased the import dependency of Nepal. Thus, the study shows that the energy security situation of Nepal has been degrading from 2005/2006 to 2016/2017.

In the scenario analysis, the final energy demand in the BAU Scenario is 824.79 PJ, while in the ES Scenario is 519 PJ in the end year 2030. The final energy demand in the ES Scenario is less than that in the BAU Scenario due to the penetration of energy efficient technologies. In the BAU Scenario, the energy index follows a declining trend throughout the study period, while in the ES Scenario, the energy security index followed a declining trend until 2018/2019. There is improvement in energy security index from 2019/2020 which can be attributed to the increase in electricity generation capacity that increases self-sufficiency and decreases electricity deficit, which helps strengthen the energy security status of the country. The value of energy security index follows an increasing trend from the first scenario year to the end year 2029/2030 in the ES Scenario.

There is a need to incorporate energy security in the national energy policy of Nepal with more emphasis on diversification of primary energy sources (other than traditional resources), reduction on import dependency especially on fossil fuel resources, strategic fuel reserves, diversification of energy resources uses in different energy consumption sectors such as transport, industries, increase in energy supply from renewable energy sources, etc. Indigenous electricity generation from hydropower sources and renewables such as solar, wind, etc. is the most promising way of ensuring energy security and sustainability issues of Nepal. On the demand side, energy management by the use of energy efficient technologies could lessen the total energy consumption and hence decrease the energy demand which could strengthen the energy security of Nepal inthe long term.

The energy security index changes with the change in normalizing and weighting methods. Nonetheless, the energy security indicators help to provide a better understanding of Nepal's ES Scenario for further analysis and improvement, while the energy security index helps to present and understand the overall energy security trend. The calculated energy security index can be useful in benchmarking and tracking the progress of the energy security situation of Nepal in the coming years.

REFERENCES

ADB. (2017). *Nepal Energy Sector Assessment, Strategy and Road Map*. Philippines: Asian Development Bank.

AEPC. (2008). *Solar and Wind Energy Resource Assessment in Nepal (SWERA)*. Lalitpur: Alternative Energy Promotion Center.

Ang, B., Choong, W., & Ng, T. (2015). Energy security: Definitions, dimensions and indexes. *Renewable and Sustainable Energy Reviews, 42*, 1077–1093.

APERC. (2007). A Quest for Energy Security in the 21st Century, Resources and Constraints. Institute of Energy Economics, Japan: Asia Pacific Energy Research Centre.

Bajracharya, T. R., & & Darlamee, S. (2020) A Research on: Energy Security, Reliability and Development, Centre for Applied Research and Development (CARD), Institute of Engineering, Tribhuvan University.

CBS. (2012). *National Population and Housing Census 2011*. Kathmandu, Nepal: Government of Nepal, National Planning Commission Secretariat, Central Bureau of Statistics.

Cherp, A., & Jewell, J. (2011). Measuring energy security: From universal indicators to contextualized frameworks. *The Routledge Handbook of Energy*, 330–355.

Child, D. (1990). *The Essentials of Factor Analysis*, (2nd, Ed.).New York, US: Casel Education.

Chuang, M., & Ma, H. (2013). An assessment of Taiwan's energy policy using multidimensional energy security indicators. *Renewable and Sustainable Energy Reviews*, *17*, 301–311.

Darlamee, S. (2019) Energy Security Assessment of Nepal for the period 2005-2030, M.Sc. Thesis, Department of Mechanical Engineering, Pulchowk Campus, Institute of Engineering, Tribhuvan University, Nepal.

Dogbegah, R., Owusu-Manu, D., & Omoteso, K. (2011). A principal component analysis of project management competencies for the Ghanaian construction industry. *Australasian Journal of Construction Economics and Building*, *11*(1), 26–40.

Doukas, H., Papadopoulou, A., Savvakis, N., Tsoutsos, T., & Psarras, J. (2012). Assessing energy sustainability of rural communities using principal component analysis. *Renewable and Sustainable Energy Reviews*, 1949–1957.

EPA.(2019,0506).*United States Environmental Protection Agency.*Retrieved from https://www.epa.gov: https://www.epa.gov/ghgemissions/understanding-global-warming-potentials

GIZ/NEEP. (2016). *Promoting Cogeneration in Nepal: Adding a New Source to the Electricity Mix*. Kathmandu, Nepal: Deutsche Gesellschaft fürInternationaleZusammenarbeit (GIZ) GmbH/Nepal Energy Efficiency Programme.

Gnansounou, E. (2008). Assessing the energy vulnerability: Case of industrialised countries. *Energy Policy*, *36*, 3734–3744.

Gonzalo, E. F., José María, M., & González, E. S. (2013). RES and risk: Renewable energy's contribution to energy security. A portfolio-based approach. *Renewable and Sustainable Energy Reviews*, *26*, 549–559.

IAEA. (2005). *Energy Indicators for Sustainable Development: Guidelines and Methodologies*. International Atomic Energy Agency.

IEA. (2007). *Energy Security and Climate Change Assessing Interactions*. Paris: International Energy Agency.

IEA. (2008).*Energy Technology Perspectives,2008:Scenarios and Strategies to 2050*. Paris: International Energy Agency.

IEA. (2001). *Toward a Sustainable Energy Future*. Paris: International Energy Agency.

Kruyt, B., Vuuren, D., Vries, H., & Groenenberg, H. (2009). Indicators for energy security. *Energy Policy*, 2166–2181.

Malla, S. (2014). Assessment of mobility and its impact on energy use and air pollution in Nepal. *Energy*,1–12.

Månsson, A., Johansson, B., & Nilsson, L. J. (2014). Assessing energy security: An overview of commonly used methodologies. *Energy*, *17*, 1–14.

Martchamadol, J., & Kumar, S. (2013). An aggregated energy security performance indicator. *Applied Energy*, 103, 653–670.

Martchamadol, J., & Kumar, S. (2012). Thailand's energy security indicators. *Renew Sustain Energy Rev2012*,103–122.

MoF. (2017). *Economic Survey 2016/17*. Nepal: Ministry of Finance.

MoPE. (2017). *Biomass Energy Strategy 2017*.Nepal: Ministry of Population and Environment.

Narula, K., & Reddy, B. S. (2016). A SES (sustainable energy security) index for developing countries. *Energy*, *94*, 326–343.

NEA. (2017). *A Year in Review 2016/17*. Nepal Electricity Authority.

NOC. (2017). *Annual Report 2016/17*. Nepal Oil Corporation.

NOC. (2016). *Prabhat 2073*. Nepal Oil Corporation.

NPC. (2015). *Sustainable Development Goals 2016-2030, National (Preliminary) Report*. Government of Nepal, National Planning Commission.

NPC. (2017). *Nepal's Sustainable Development Goals,Status and Roadmap: 2016-2030.* Singhadurbar, Kathmandu: Government of Nepal, National Planning Commission.

NPC. (2018).Universalizing Clean Energy in Nepal, *A Plan for Sustainable Distributed Generation and Grid Access to All by 2022.* Singhadurbar, Kathmandu: Government of Nepal, National Planning Commission.

OECD. (2008). *Handbook on Constructing Composite Indicators: Methodology and User Guide.* Organisation for Economic Cooperation and Development.

Phdungsilp, A. (2010). Assessing energy security performance in Thailand under different scenarios and policy implications. *Energy Procedia, 79,* 982–987.

RStudio Team. (2015). *RStudio: Integrated Development for R.* Boston, MA: RStudio, Inc.

Scheepers, M., Seebregts, A., Jong, J., & Maters, H. (2006). *EU Standards for Energy Security of Supply.* The Netherlands: ECN.

Shakya, S. (2014). *Energy Efficiency Improvement Potential of Nepal.* Pulchowk, Lalitpur: Center for Energy Studies (CES), Institute of Engineering, Tribhuvan University.

Shakya, S. R., & Shrestha, R. M. (2011). Transport sector electrification in a hydropower resource rich developing country: Energy security, environmental and climate change co-benefits. *Energy for Sustainable Development, 15*(2), 147–159.

Shin, J., W.-S. Shin, & Lee, C. (2013). An energy security management model using quality function deployment and system dynamics. *Energy Policy, 54,* 72–86.

Shrestha, S., & Nakarmi, A. (2015). Demand side management for electricity in Nepal: Need analysis using LEAP modeling framework. *Proceedings of IOE Graduate Conference,* 242–251.

Sovacool, B. K. (2013). An international assessment of energy security performance. *Ecological Economics, 88,* 148–158.

Sovacool, B. K., & Ren, J. (2014). Quantifying, measuring, and strategizing energy security: Determining the most meaningful dimensions and metrics. *Energy, 76,* 838–849.

Timilsina, G., Sapkota, P., & Steinbuks, J. (2018). How Much Has Nepal Lost in the Last Decade Due to Load Shedding? An Economic Assessment Using a CGE Model.World Bank Working Paper.

Tongsopit, S., Kittner, N., Chang, Y., Aksornkij, A., & Wangjiraniran, W. (2016). Energy security in ASEAN: A quantitative approach for sustainable energy policy. *Energy Policy, 90,* 60–72.

Vivoda, V. (2009). Diversification of oil import sources and energy security: a key strategy or elusive objective. *Energy Policy,* 4615–4623.

Wang, Q., & Zhou, K. (2017). A framework for evaluating global national energy security. *Applied Energy, 188,* 19–31.

WEC. (2018). *World Energy Trilemma Index.* World Energy Council.

WECS. (2010). *Energy Sector Synopsis Report 2010.* Kathmandu, Nepal: Water and Energy Commission Secretariat.

WECS. (2013a). *Nepal's Energy Sector Vision 2050 (Draft).* Kathmandu, Nepal: Water and Energy Commission Secretariat.

WECS. (2013b). *National Energy Strategy of Nepal (Draft).* Kathmandu, Nepal: Water and Energy Commission Secretariat.

WECS. (2013c). *National Survey of Energy Consumption and Supply Situation in Nepal.* Kathmandu, Nepal: Water and Energy Commission Secretariat.

World Bank. (2005). *Energy Security Issues, Briefing Paper.* The World Bank.

World Bank. (2014). RISE, Readiness for Investment in Sustainable Energy, *A tool for policymakers.* International Bank for Reconstruction and Development. The World Bank.

Wu, G., Liu, L., Han, Z., & Wei, M.(2012). Climate protection and China's energy security: win-win or tradeoff. *Applied Energy, 97,* 157–163.

Yao, L., & Chang, Y. (2014). Energy security in China: A quantitative analysis and policy implications. Energy Policy, *67,* 595–604.

APPENDIX OF INDICATORS FORMULATION

S.N.	Indicators	Symbol	Formula	Unit

Availability Dimension

1. Shannon-Weiner Index — AVL1

$$D = \sum (p_i \ln p_i)$$

D = Shannon's Diversity Index
P_i = share of PES I in TPES

2. Net Energy Import Dependency — AVL2

$$NEID = \frac{\sum_i (m_i p_i \ln p_i)}{\sum_i (p_i \ln p_i)}$$

M_i = share of net imports in PES of source i
P_i = share of PES I in TPES

3. Net Oil Import Dependency — AVL3

$$NOID = \frac{Net\ Oil\ Imports}{Oil\ Primary\ Energy\ Demand} \times \frac{Oil\ PED}{Total\ PED}$$

4. No. of days of stocks of petroleum products — AVL4 — No. of days of stock=(Storage capacity/Annual sales) * 365 — days

5. No. of days of stocks of LPG — AVL5 — No. of days of stock = (Storage capacity/Annual sales) * 365 — days

6. Share of domestic production — AVL6 — Domestic production (hydro+ biomass+ alternatives) /Total PES — %

7. Annual electricity deficit — AVL7 — (Peak Demand-Installed capacity) /Installed capacity — %

8. Electricity Consumption per Capita — AVL8 — Total Electricity Consumption/ Population — kWh/capita

Affordability Dimension

9. Value of Petroleum Products Imports per Unit of GDP — AFF1 — Expenditure on imports of petroleum products/GDP

Efficiency Dimension

10. Total Final Energy Consumption per Capita	EFF1	Final Energy Consumption/ Population	GJ/capita
11. Final Energy Intensity	EFF2	Final Energy Consumption/GDP at 2000/01 Prices	GJ/1000NRs.
12. Oil Consumption per Capita	EFF3	Total Petroleum Product Consumption/Population	GJ/capita
13. Oil consumption per unit of GDP	EFF4	Petroleum Products Consumption/ GDP at 2000/01 Prices	GJ/1000NRs.
14. Residential Energy Intensity	EFF5	Residential Energy Consumption/ Total Population	GJ/HH
15. Industrial Energy Intensity	EFF6	Industrial Energy Consumption/ Value Addition by Industrial Sector	GJ/1000NRs.
16. Commercial Energy Intensity	EFF7	Commercial Energy Consumption/Value Addition by Industrial Sector	GJ/1000NRs.
17. Agricultural Energy Intensity	EFF8	Agricultural Energy Consumption/ Value Addition by Industrial Sector	GJ/1000NRs.
18. Transport Energy Intensity	EFF9	Transport Sector Energy Consumption/km Travelled by Passenger and Freight Vehicles	GJ/pkm

Acceptability Dimension

19. Non-Carbon Fuel Portfolio	ACP1	*NCFP*	%

$$NCFP = \frac{Hydro\ PED + Nuclear\ PED + NRE\ PED}{TPED}$$

ACKNOWLEDGMENT

We would such as to acknowledge the support received from the different persons and institutions to complete this study. First, we would such as to give our sincere thanks to the Centre for Applied Research and Development (CARD) of the Institute of Engineering, Tribhuvan University. The support was received from a Master of Science in Energy for Sustainable Social Development (MSESSD) program under Institute of Engineering (IOE) and Norwegian University of Science and Technology (NTNU) joint project funded by the Energy and Petroleum Program of the Norwegian Government.

8 Pakistan's Energy Transformation Pathway and Environmental Sustainability

Naila Saleh

CONTENTS

8.1 Introduction 175
8.2 Overview of Power Sector Challenges 177
    8.2.1 Energy Access Gap.. 178
    8.2.2 Import Dependency 178
    8.2.3 Circular Debt 180
    8.2.4 Transmission and Distribution Losses 180
8.3 The Power Sector Transformation 181
    8.3.1 Historical Background.. 181
    8.3.2 CPEC and the Evolving Energy Mix... 184
8.4 Socio-Economic and Environmental Challenges 186
8.5 Slow Renewable Energy Uptake.. 188
8.6 Conclusion 190
References.. 191

8.1 INTRODUCTION

Energy is a vital strategic commodity for modern societies. Since the advent of the industrial revolution, wider developments have been accomplished through the utilization of different forms of energy. The extensive reliance on fossil fuels, however, has created its own catastrophic threats with climate change emerging as one of its defining challenges. The pursuit for a climate-safe energy system, hence, has prompted the quest for sustainable supplies worldwide. Today an increasing number of countries have already mainstreamed more than 20% of solar and wind in their power sectors (Renewables 2019 Global Status). These transitioning economies encompass many South Asian counties including China which accounted for 32% of global wind and 35% of global solar capacity, India ranking sixth in global solar-based capacity and fourth in terms of installed wind energy capacity, and Bangladesh remaining an emerging hotspot for the global off-grid renewable installations (*Asian Power*, 2019).

Where does Pakistan stand amid this transition? The generation capacity in Pakistan has been undergoing significant growth in the past few years. Massive projects of a cumulative capacity of 14,184 MW have already been incorporated into the supply side during 2015–2019 (NEPRA, 2019). Further, in the next five years, Pakistan plans an additional 27 GW of capacity in the power sector (NEPRA, 2018a). However, in sharp contrast to the global trajectory—galloping ahead to embrace wind and solar energy—the recent power sector transformation efforts in the country have ushered in a new phase for coal-fired power plants where the recently added and planned capacities are largely dominated by outdated and expensive coal. A lion's share of these added and planned capacities are coal-based—4.6 GW of added capacity (by December 2019) and 10 GW of planned capacity (see Table 8.1). These developments have increased the share of coal in the power mix to 19% of the total generation mix in 2019, compared to less than 1% in 2015 (NEPRA, 2019).

For a country already confronted with alarming environmental threats, this new energy mix is likely to significantly increase greenhouse gas (GHG) emissions and exacerbate the impact of climate change. Pakistan falls among the most vulnerable geographic regions to climate change (ADB, 2017a). Erratic weather patterns, heat waves, floods, and other extreme events have become a regular phenomenon. Increasing evidence showcases the health damages and productivity losses arising out of these disturbing changes (Muthukumura et al., 2018). Climate change has also aggravated the water crisis in the country, wherein Pakistan is expected to be one of the most water-stressed countries by 2040 (WRI, 2015). Not to overlook the level of different pollutants in most urban cities far exceeding the national and international

TABLE 8.1

Planned Capacity Additions in Pakistan's Power Sector (up to 2025)

Technology	2018-19	2019-20	2020-21	2021-22	2022-23	2023-24	2024-25	Total
Domestic Coal	660	-	2310	660	1320	-	-	4950
Imported Coal	660	660	163	2280	1320	-	-	5083
Oil	-	-	-	-	-	-	-	
Gas/Liquefied Natural Gas (LNG)	800	463	-	-	-	-	-	1263
Wind	-	250	-	750	-	-	-	1000
Solar	800	100	300	-	-	-	-	1200
Bagasse/ Biomass	348	389	60	-	-	-	-	797
Hydro	78	102	261	2984	1040	3786	1865	10116
Nuclear	-	-	1100	1100	-	-	-	2200
Import	-	-	-	1000	-	-	-	1000
Yearly Addition	3345	1964	4244	8774	3680	3786	1865	27658

Source: Estimated from NEPRA, 2018a.

thresholds where the energy sector remains one of the largest contributors to air pollution (Sanchez et al., 2014; USAID, 2012).

In addition to the environmental sustainability challenges, dependence on imported fossils—earlier oil and recently Re-Gasified Liquefied Natural Gas (**RLNG**) and coal— is another emerging challenge in the broader context of already economically unsustainable power sector performance. The share of fast-rising imports and currency devaluation have significant implications for the energy sector. Currently, imported fuel dependence in the power sector of the country stands at 55% (Shibli, 2018). The chronic deficits and inter-linked quasi-fiscal losses in the power sector—arising in the backdrop of a shift toward expensive thermal sources and currency devaluation—have undermined incentives for investment in new capacity to address supply shortages. Wherein Pakistan holds one of the largest populations living off the grid (around 50 million still have no access to the grid)—the increase in overall generation cost in parallel to the quasi-fiscal losses in the power sector will remain a major challenge to affordable and countrywide energy accessibility. So where availability of sufficient and affordable energy is critical to tackle poverty and improve living standards, the prevailing unsustainable trajectory will continue to be responsible for the poor public service in terms of energy provision.

Despite the stated wide-scale energy sector challenges, the transition initiatives in Pakistan have largely been based on traditional interventions which are principally focused on plugging the energy access gap and diversification of fuel resources. The new energy pathway has further deepened the country's carbon dependence—particularly the alarming development of a coal- powered industry from scratch. Expansion of power sector generation by fossil fuels will increase GHG emissions which will not only result in environmental degradation, but also delay the achievement of sustainable development goals (SDGs) making it more difficult (if not impossible) to meet the emissions' targets under "Intended Nationally Determined Contribution (INDC)" of the Paris Agreement. The undergoing trajectory hence makes it really intriguing to observe the underlying factors contributing to the untimely "Coal Phase-In", and not otherwise tapping into the immense, clean indigenous energy sources which the country holds in abundance. Understanding the complexities involved in changing the status quo and transitioning to a desirable energy system hence are critical.

This chapter walks the readers through Pakistan's energy profile. It provides key insight on the nature and magnitude of multi-dimensional challenges in the power sector, recent as well as past power sector transformation initiatives, sustainability challenges of the new energy pathway, and finally an overview of major obstacles slowing down decarbonization of the power sector and the renewable energy (RE) uptake.

8.2 OVERVIEW OF POWER SECTOR CHALLENGES

Pakistan is one of the growing economies in the South Asian region. However being chronically undercapitalized and characterized by insufficient generation, import reliance, deteriorating equipment, high system losses, inefficient financial performance—the country has never enjoyed a stable energy sector. The following sections briefly overview major challenges facing the country's power sector.

8.2.1 ENERGY ACCESS GAP

Pakistan is home to one of the biggest "energy poor" people—holding the fourth largest unserved population globally, i.e., translating to around 50 million people (IEA, IRENA, and WB, 2019). At the same time, deficits in energy supply are an additional manifestation of energy poverty where the magnitude of the challenge is significantly apparent in Pakistan. The country has been facing some of the worst global electrical blackouts, where the gap between supply and demand kept widening from 2006 onwards. The shortfall reached 26% of the country's total demand in 2013, leading to prolonged load shedding hours, i.e., 6–14 hours a day (Zhang, 2018). These shortages largely stemmed from several factors including a consistent increase in energy demand, insufficient investment in generation, aging transmission and distribution infrastructure, the interlinked chronic line losses, and substantial arrears in the value chain of energy supply. Figure 8.1 shows the power deficit registered during peak demand hours from 2006 to 2018.

8.2.2 IMPORT DEPENDENCY

Energy security has posed challenges to both energy-rich as well as import-dependent countries where the concerns of the former are linked to maintenance of export revenues, and the latter on the implication of their dependency on imported fuels. Pakistan has also been historically reliant on imported fuels for meeting its energy needs. This reliance has collectively contributed to serious energy insecurity challenges and poor financial position of energy companies.

In terms of electricity generation, the country relied largely on hydropower for meeting the electric power needs in its early years. However, thermal power plants—initially built as a backup for mitigating the seasonal fluctuations of hydropower generation capacity—kept growing at a striking speed where hydropower's share in terms of installed generation capacity fell from around 67% in 1985 to 24% in 2018 (NEPRA, 2006-19). Failure to mobilize the funds for hydropower projects—which

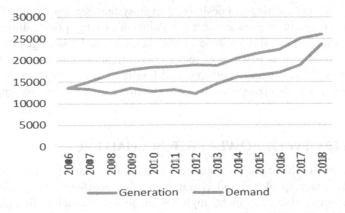

FIGURE 8.1 Power deficit in demand (peak hours) and supply (MW). (Source: Estimated from NEPRA, 2006-19.)

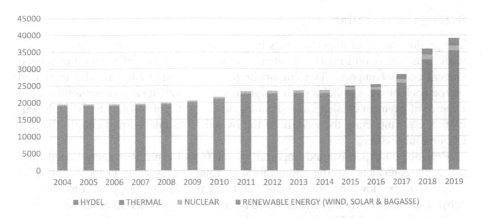

FIGURE 8.2 Electricity generation installed capacity by source (2006-2019). (Source: Estimated from NEPRA, 2006-19b.)

entail significant upfront capital cost compared to thermal sources—remained the leading factor for this altered landscape. Around two-thirds of existing electricity generation comes from thermal fuels—62.1% of the energy mix. Whereas hydro, nuclear, and renewable makes 25.8%, 8.2%, and 3.9% of the generation mix, respectively (Economic Survey of Pakistan, 2019). Figure 8.2 shows the source-wise electricity generation capacity of Pakistan since 2004.

Oil and gas dominate the energy supplies of Pakistan; however, they account for a minimal share of indigenous production—in the case of oil only around 17% of the country's needs (Zhang, 2018). Although until recently Pakistan had sizeable reserves of natural gas, its intensified use in the wake of no large new discoveries has contributed to the continued widening gap between its supply and demand. As per the forecasts, this shortage will reach an estimated 3,399 million cubic feet (mmcfd) per day by 2019–2020 (OGRA, 2018). Figure 8.3 shows the projected demand-supply gap of natural gas.

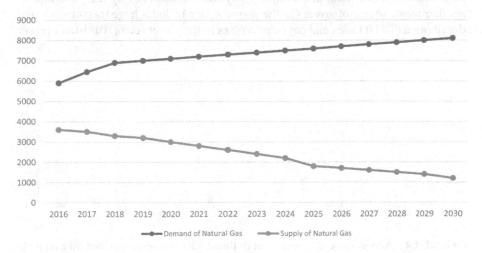

FIGURE 8.3 Demand-supply gap of natural gas (mmcfd).

8.2.3 CIRCULAR DEBT

The predominance of imported fuels in the energy mix has increased the country's trade bills, where oil import bills alone account for a quarter of the total estimated current account imports. This import-driven growth model has rapidly increased on the back of weak policies posing significant macroeconomic vulnerabilities, the balance of payment challenges, and contributing disproportionately to the high generation cost impacting the welfare of the overall consumers—particularly the less resourceful sections.

Reliance on expensive and imported major fuels for meeting energy needs significantly contributed to the appreciation of final energy costs in the country. To keep the prices low, the government has been providing energy to consumers at subsidized rates. Currently, end-users consuming 300 units (kWh) or below—which accounts for around 70% of all household consumers—are provided subsidies and insulated from energy cost appreciation. The government's policy to underprice energy requires large subsidies from the budget. However, under-budgeting of these subsidies, as well as delays in their disbursement, has been contributing to persistent inter-sectoral debts in the supply chain contributing to the vicious accumulation of huge arrears in the power sector denoted by "circular debt"—the quasi-fiscal losses and cash flow shortfall incurred in the power sector from the non-payment of obligations by consumers, distribution companies and the government. This debt has long impaired the power sector of Pakistan, having a knock-on effect on the entire electricity supply chain in terms of supply cuts at every level of the energy chain and the resultant idled generation capacity and power cuts to consumers, thus imperiling the creditworthiness of public and private generation companies. Figure 8.4 shows the stock of annual circular debt registered during 2006–2019.

8.2.4 TRANSMISSION AND DISTRIBUTION LOSSES

In addition to insufficient and untimely payment of subsidies by the government, another major source of arrears in the power sector includes huge transmission and distribution (T&D) losses and payment arrears in the power sector. Pakistan's power

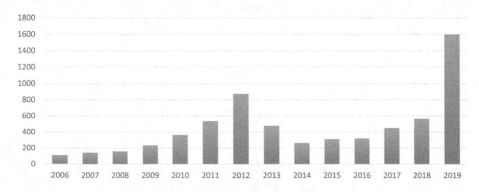

FIGURE 8.4 Annual stock of circular debt (billion PKR). (Source: Estimated from multiple sources including USAID, 2013; Illangovan & Francisco, 2017; Saleh, 2019)

TABLE 8.2
Transmission and Distribution (T&D) Losses

DISCO	2013-14	2014-15	2015-16	2016-17	2017-18
IESCO	9.46	9.41	9.10	9.02	9.13
PESCO	33.5	34.8	33.8	32.6	38.1
GEPCO	10.97	10.72	10.58	10.24	10.01
FESCO	11.3	11	10.2	10.6	10.5
LESCO	13.4	14.1	13.9	13.8	13.8
MEPCO	17.5	16.7	16.4	16.9	16.6
QESCO	28.3	24.4	23.8	23.1	22.4
SEPCO	38.56	38.29	37.72	37.8	36.7
HESCO	26.46	27.1	26.5	30.8	29.8
K-ELECTRIC	25.30	23.69	22.24	21.71	20.4

Source: NEPRA, 2018b.

distribution losses are one of the highest in South Asian countries. Against the international reference point of 10% threshold, these losses on average exceed 20% in the case of Pakistan (NEPRA, 2018b). In the fiscal year (FY) 2018 alone, they amounted to around 187 billion PKR (Faraz, 2018). Energy theft, inadequate size of conductors, lengthy distribution lines, and installation of distribution transformers away from load centers included a few underlying factors responsible for these technical losses. Further, poor recovery of revenues from end-users by distribution companies (DISCOs) further exacerbate the financial shortfalls costing additional losses to the national exchequer—around 78 billion PKR during the stated period (NEPRA, 2018b). Table 8.2 illustrates the T&D losses recorded across the 10 distribution companies from 2014 to 2018.

How did the power sector become ostensibly unsustainable and reliant on imported expensive fuels in the first place? The subsequent section takes stock of Pakistan's power sector transformation experiences in its early history and the resulting dramatic implications on the power landscape (as discussed in this section) in terms of the expensive energy mix, circular debt, and the inter-linked chronic power crisis.

8.3   THE POWER SECTOR TRANSFORMATION

8.3.1   HISTORICAL BACKGROUND

The early phase of power sector reform in Pakistan could be traced back to the mid-1990s' electricity crisis. At the time, the country had a vertically integrated and autonomous power and water utility statutory body—Water and Power Development Authority (WAPDA)— to develop and coordinate schemes in the water and power sectors. However, rising energy prices, rampant corruption, poor financial performance, and power shortfalls led to the realization that the power system was becoming too large to be handled alone by a single integrated body. Eventually, a momentum for introducing reform package was mobilized to improve efficiency via "unbundling, regulation, and privatization" of the existing generation and distribution.

The reform began with the unbundling of WAPDA in 1992, wherein the power wing was bifurcated into several state-owned enterprises respectively for generation, transmission, and distribution (Bacon, 2019). Following it was a new "National Power Policy" which was formulated in 1994—opening generation to private investment. In parallel, a new state-owned institution Private Power and Infrastructure Board (PPIB) was established, principally responsible for providing guidance and implementing agreements related to private power projects. Following it, in 1997, a regulatory entity National Electric Power Regulatory Authority (NEPRA) was established with the major task to ensure "safe, reliable, efficient and affordable electric power to the electricity consumers of Pakistan, and to facilitate the transition from a protected monopoly service structure to a competitive environment" (Khalid et al., 2016). NEPRA was also entrusted with the task to grant licenses and set tariffs for both generation and distribution. Whereas for overseeing the unbundling and privatization of WAPDA, another institution, Pakistan Electric Power Company Limited (PEPCO), was established in 1998. Hence, PEPCO had to design a new bureaucratic structure aiding the transition to a new commercially viable model and managing the thermal generation plants formerly overseen by WAPDA. However, hydropower generation remained with WAPDA.

The 1994 power policy was replaced by a new national power policy in 2002. Responsibility for a smaller generation—i.e., less than 50 MW—under the new policy was placed with provincial governments. Also, the fuel supply guarantees under the previous policy were removed. Further, a single buyer model has been in operation since the sector restructuring, with the Central Power Purchasing Agency (CPPA)—fully separated from National Transmission and Dispatch Company (NTDC) which transmits power between generators and the DISCOs. CPPA acts as the power market operator pooling the electricity purchased from all the generation companies (GENCOs), independent power producers (IPPs), WAPDA hydro, and then selling it to the DISCOs which further distribute it among end-users.

In order to attract private investment in renewables, i.e., small hydro, wind, and solar Photovoltaic (PV) plants, an exclusive organization Alternative Energy Development Board (AEDB) was established in 2003. A new RE policy, "Policy for Development of Renewable Energy for Power Generation" was issued during the same year by AEDB. Further, new power policies kept replacing old policies at different intervals of time during this entire period. As of 2019, Pakistan had 10 distribution companies, 4 GENCOs, 36 private IPPs, 3 nuclear power plant companies, a transmission, and system operator (National Transmission and Dispatch Company, NTDC) and an electricity market operator, Central Power Purchase Agency (CPPA-G) (ADB, 2019). Figure 8.5 shows the flowchart of structural and policy reforms undertaken since 1992.

Although initially the reform plan also envisioned gradual privatization of distribution, it was later on shelved—except in the case of one distribution utility Karachi Electric (KESCO) which was privatized in 2005. The reform also envisioned the transformation of the existing single buyer model into an open-access competitive market, yet to be materialized.

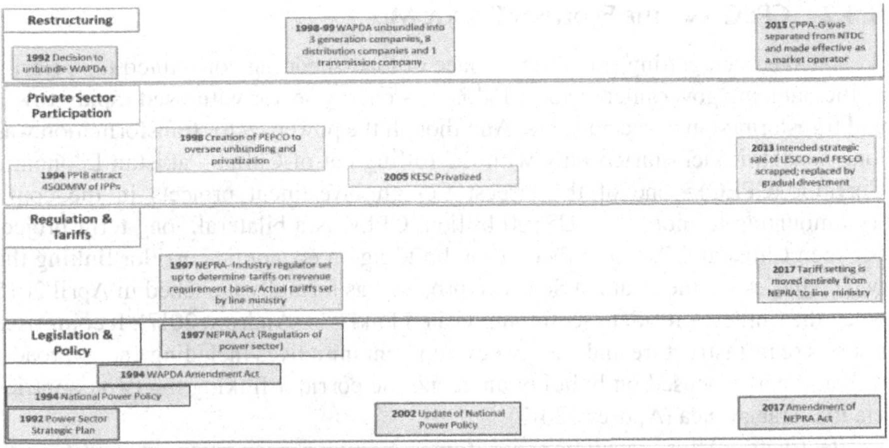

FIGURE 8.5   Restructuring and policy changes in the power sector of Pakistan.

Although the reforms undertaken in the early phase of Pakistan's history altered the power sector landscape—unbundling the power sector and opening up generation to private sector participation—for much of the period since the introduction of these reforms, the power sector continued to suffer from multi-dimensional challenges. The cost-plus tariff regime introduced under the 1994 policy guaranteed fixed returns to the investors over the life cycle of their project, placing the onus of currency risks and fuel supply with the government (Bacon, 2019). Additionally, owing to the government's urgent need to address the power crisis, the policy favored developers bringing their plants online fast, and so most of the thermal plants incorporated during this period were oil-based. Eventually, these reforms brought a substantial transformation of Pakistan's power generation mix and proved its inception point of heavy reliance on imported fuel generation mix which remained the face of the generation addition for over three decades.

All these multi-dimensional changes facing the energy sector contributed to chronic financial deficits in the power sector, and under-investment in generation with electricity shortfalls emerged as its largest consequence. The electricity shortages and prolonged load shedding in the country also had large-scale social, economic, and political implications. The shortages cost the economy an estimated 2% of gross domestic product (GDP) annually (Economic Survey of Pakistan, 2010). Public protests and demonstrations against the power outages necessitated a reform program for overcoming the crisis. Among many other commitments, the Nawaz Sharif government was swept into power in the 2013 elections partly due to its pledges of tackling the energy problems that its predecessors had failed to address. Overcoming the severe energy crisis became the top priority of the outgoing government. The new government hence wasted little time in engaging China for investment in Pakistan's power sector. In parallel, China was receptive to the investment plan largely because of its aligned interest in the construction of a bilateral economic corridor between the two countries. Finally, mutual interest laid the groundwork for the launching of China–Pakistan Economic Corridor (CPEC) and substantial transformation in Pakistan's power sector.

8.3.2 CPEC AND THE EVOLVING ENERGY MIX

As stated earlier, getting rid of the menace of load-shedding constituted top priority of the outgoing government and so Pakistan's energy sector witnessed rapid growth and transformation in recent years. And though the power sector transformation was long sought, it kick-started only with the rolling out of China–Pakistan Economic Corridor (CPEC)—one of the largest foreign investment projects in the country amounting to more than US$60 billion. CPEC is a bilateral, long-term project between China and Pakistan focused on building an economic corridor linking the two countries via the Arabian Sea. The project was formally launched in April 2015 when the Chinese president Xi Jinping visited Pakistan (Andrew, 2017). It comprises a series of infrastructure and energy development initiatives including energy, roads, rail, and ports focused on building an economic corridor linking the two countries via the Arabian Sea (Andrew, 2017).

Electricity sector generation constitutes a major proportion of CPEC, i.e., accounting for around 55% of the total project value and 45% of the country's total installed capacity at that time, i.e., June 2015. The project was planned into four stages where keeping in perspective the government's priority of easing the energy shortages, most energy-related projects were put under the first stage, i.e., "Early Harvest 2015–2019" and have already reached their successful conclusion.

While the main aim of the Nawaz government was to enhance generation capacity, the recent currency devaluation with similar past trends also stressed the need for reducing dependence on expensive fuel imports—a major drain on the country's foreign exchange reserves. Switching to cheap alternate sources and phasing out oil constituted the cornerstone of the power sector transformation plan. Compared to other fossil fuels, Pakistan holds huge coal reserves estimated at over 186 billion tons (Economic Survey of Pakistan, 2018). The country has long sought to improve energy security and reduce reliance on imported fuels by developing its vast coal reserves for power generation. Successive power sector policies in the country have emphasized the role of coal and the development of indigenous resources for transforming the country's power generation fuel mix. However, this goal remained largely unmet due to financing, infrastructure, and technical limitations. For instance, many companies were reluctant to invest in Thar coal partly because of the poor quality coal and high cost of mining. Moreover, the financial restrictions for the development of coal-fired generation by many international financial institutions such as the World Bank remained as other impediments. This quest was finally dovetailed with CPEC which substantially spurred the share of coal in the evolving energy mix, while aiding both phasing out of oil and accommodating for falling domestic gas production.

During 2015–2019, massive power projects of around 14,184 MW cumulative capacity were incorporated (NEPRA, 2019). Alarmingly, three-quarters of this new generation capacity under CPEC is coming from coal-fired power plants—reaching 25% of power mix as of lately. Hence the evolving energy mix will expand the coal's share of the country's generation capacity from 0.1% of power generation in 2015, i.e., before the launch of the project to nearly 20 % by 2025 (NEPRA, 2017). Also, it is very important to note here that most of the coal-based additions are comping for imported coal—and not the otherwise anticipated larger utilization of indigenous

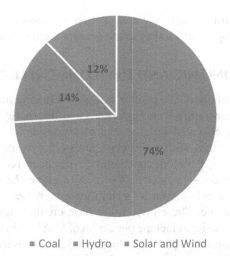

■ Coal ■ Hydro ■ Solar and Wind

FIGURE 8.6 Generation capacity of CPEC power plants by fuel. (Source: CPEC Secretariat,.)

coal. Figure 8.6 shows distribution of CPEC projects in terms of technologies, whereas Figure 8.7 shows the fuel-wise comparison of thermal power generation mix from 2014 to 2019.

So, although the country has planned a prudent phase-out of the expensive oil-fired power generation, the import dependency has further deepened recently owing to its replacement by imports of alternate fuels. Coal has played a negligible role in Pakistan's power sector historically. The recent drastic shift toward it reflects the country's longstanding goal of diversifying the energy mix and moving toward cheap fossil fuels for power generation. These developments also reflect the conviction that increasing the role of coal would reduce the energy cost and the country's

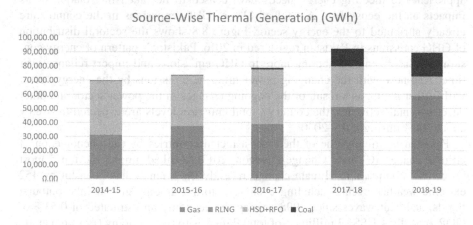

FIGURE 8.7 Fuel-wise comparison of thermal power generation 2014–2019. (Source: Estimated from NEPRA, (2006-19)

import bill. Against this context, the next section previews the large-scale implications of the undergoing developments in the country's power sector.

8.4 SOCIO-ECONOMIC AND ENVIRONMENTAL CHALLENGES

Despite the adoption of two-fold transition agenda—reducing import reliance and the final cost of energy generation—the performance of the energy sector remains fragile and has further deteriorated with the recent substantial new additions. While the challenge of power outages has largely been addressed corresponding to the substantial new additions in a generation; nonetheless, it has occurred at the cost of a huge increase in energy prices—by nearly 24% in 2018 alone (*Express Tribune*, 2018). The deterioration is largely driven by a significant increase in generation capacity during a short timeframe. These price hikes have created significant affordability challenges. Given the national income per capita of US$1,497 and the global average of over 3,000 kWh consumed per individual each year, on average around 20% of a Pakistani individual's income is spent on electricity consumption, compared to 1.1% for the US, 2.3% for Germany, and 3.1% for China (Raza, 2019).

Additionally, although reliance on imported oils has narrowed down recently, nevertheless dependence on external resources continues owing to its replacement by alternate imported fuels and not otherwise indigenous resources. For instance, in the case of oil, its share was successfully brought down from was 43.5% of the power mix in 1992 to 31.2% in 2018. Against this decline, the share of imported coal, LNG and gas is continuously on the rise. The share of imported LNG registered 8% growth during the past three years, i.e. from 0.7% in FY2015 to 8.7% in FY2018. In the case of coal, it registered a 25% increase during the same period—12.7% alone in FY2018. So although the energy mix is changing, it remains heavily skewed toward fossil fuels. Hence, the quest for energy security in the power sector remains elusive owing to the imprudent reforms—correcting one distortion at the cost of others.

As the energy sector is the leading contributor to climate change globally, approaches to meeting energy needs have ceased to be taken in isolation of its impacts on the ecosystem. A total of 42% of GHG emissions in the country are already attributed to the energy sector. Figure 8.8 shows the sectoral distribution of GHG emissions in Pakistan registered in 2016. Pakistan's pattern of energy consumption has been contributing more to GHG emissions and import reliance over time. And now with the changing power mix, the emissions by the energy sector will further go up, as a result of the ongoing changes in the power sector and other developmental activities, the country's total emission levels are expected to rise by about 300% during 2015–2030.

Pakistan is already one of the most affected countries by the phenomenon of climate change (Climate Change Division, 2013). Ranked among the top 5 most vulnerable countries to climate change, recently the country has been hit by 152 extreme weather events including cyclones, storms, floods, glacial lake outburst floods, and heat waves since 1998 costing the economy an estimated of 0.53% of GDP annually—US$3.7 million—of total damages to the economy (Eckstein et al., 2020). These risks will hence remain one of the leading threats surrounding the fossil fuel- dominated energy system.

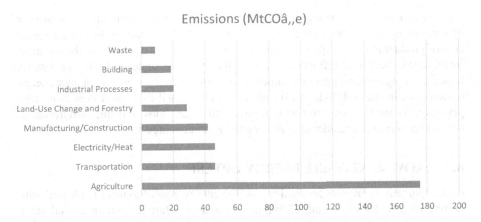

FIGURE 8.8 Sectoral share of GHG emissions in 2016. (Source: World Resources Institute Climate Analysis Indicators Tool (WRI CAIT) 2.0, WRI, 2016.)

In parallel, air pollution—again attributed largely to the energy sector—remains another leading threat to public health in the country. Though data on air quality in Pakistan is uneven, available sporadic statistics shows dangerously higher concentrations of different pollutants—especially both PM_{10} and $PM_{2.5}$. Almost every year, major cities experience extreme smog. As per a latest report of Environmental Performance Index, among a total of 180 countries Pakistan has been ranked 176 on environmental health and most alarmingly 180 on air quality (Environment Performance Index, 2020). Consequently, the country has registered some of the worst global incidences of health damages arising out of the stated ambient air quality. An estimated 135,100 deaths in the country were attributed to $PM_{2.5}$ exposures in 2015 (HEI, 2017). Additional 5000 deaths were attributed to exposure of ozone, during the same period (HEI, 2017). The age-standardized loss of healthy life expectancy (disability-adjusted life years per 100,000 people) from exposure to fine particulate matter in Pakistan also exceeds that of Bangladesh and India and is 10 times higher than the United States (HEI, 2017).

As the CO_2 content of coal features more prominently when it comes to environmental sustainability, the ruinous toll on human life is expected to further rise with the undergoing shift in power base. First, these new additions will substantially increase GHG emissions, lagging the commitments made at the United Nations Framework Convention on Climate Change under the Paris Agreement. Secondly, the increased emissions of air pollutants owing to these developments will further exacerbate air pollution and related health issues. As both coal mining and combustion for power generation require huge amounts of water, the added generation and development of the coal industry will aggravate Pakistan's water scarcity challenges.

So where on one hand the country holds one of the world's highest health burdens attributable to poor air quality, on the other hand, the failure to reflect these costs in energy prices continues to disadvantage renewables. At present, Pakistan is not characterized by substantive environmental regulations. The tariff arrangement in the country does not account for any environmental costs or damages and remains premised solely on the principle of cost recovery and a reasonable rate of return for generators. In the absence of these externalities—which are part of the overall *social* cost of producing electric power including the value of any damages to the environment, human health, or infrastructure—the total cost calculated is highly deceptive, hence masking the striking environmental advantages of the new and cleaner energy options. So, in the end, hidden externalities shelter fossil fuel generators from true price signals and the external cost of fossils (which has colossal health, environmental and economic implications) continues to be unaccounted for.

8.5 SLOW RENEWABLE ENERGY UPTAKE

Geographically located in an ideal renewable energy zone, Pakistan is blessed with more than 1500 kWh/m^2 and 5.5 Wh m^{-2} d^{-1} solar insolation, with an annual mean sunshine duration of 8–10 h d^{-1} throughout the country and around 668 W/m^2 of mean power wind density in the southern part of the country (wind speed 5–7 m s^{-1} persists in the coastal regions of Sindh and Baluchistan) (World Bank ESMAP, 2014). So renewables have huge potential in the country; yet it has a negligible share in the energy matrix—less than 5% of Pakistan's existing power generation fuel mix. Figure 8.9 shows the percentage share of installed capacity and generation mix of renewable power plants in Pakistan. Further, in the next five years, Pakistan plans 1 GW of wind and 1.2 GW of solar out of a total of 27 GW planned capacity.

The roadblocks to the uptake of variable renewable energy (VRE) are multi-dimensional in Pakistan—rooted in several factors including policy, institutional, regulatory, financial, and technological constraints. To begin with, a lack of understanding of framing the right policy for attracting investors in the renewable energy sector lies at the root of negligible VRE uptake in the country. Against a stable tariff regime for thermal plants, the unending saga of RE tariff changes and retroactive cuts have plagued Pakistan's renewable electricity sector, discouraging a large number of willing investors. Also tactlessly, the levelized tariff for most thermal projects

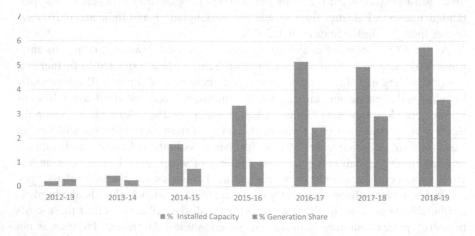

FIGURE 8.9 Percentage share renewable energy installed and generation capacity (2012–2019). (Source: Estimated from NEPRA,.)

such as coal projects (US cents 8/kWh) are more than solar levelized tariffs determined (averaging at US cents 5.25/kWh) and wind projects (averaging at US cents 4.3/kWh), hence creating uneconomic long-term obligations for potentially obsolete technology in the wake of climate change vulnerabilities facing the country (IEEFA, 2018. Tariff arrangement in Pakistan also does not account for any environmental costs or damages and remains premised on the principle of cost recovery and reasonable rate of return for generators, hence masking the striking environmental advantages of the new and cleaner energy options. Additionally, financial difficulties arising out of output curtailment—despite their "must-run" status—pose another challenge to renewable energy investors. The power market operator in Pakistan has recently reportedly curtailed zero marginal cost wind generation to zero, whilst illogically accepting offtake from coal and gas generators reliant on expensive fossil fuel imports (*Dawn*, 2019). For a nascent industry, the move once again signaled the cloying hiatuses for bankability of RE investment in the country.

Related to bottom-up transition solutions, small-scale prosumers could become key players in the energy transition drive of Pakistan owing to certain inherent forces and synergies. Pakistan is a country comprising of more than 212 million people. As stated earlier, over half of the energy demand in the country is still not met, as 46.3 million people have no access to the grid and another 144 million lack reliable access to an energy supply. The prevailing energy access gap has created an instinctive self-desire for prosumption and already distilled a momentum for alternate energy systems. Today, large sections of society hold some kind of energy backups for fulfilling their energy demands during the blackouts. Around US$2.3 billion per annum is spent by Pakistani households on alternative lighting products alone (IFC, 2015). Further, a greater share of energy consumption in the country originates from the residential sector—51% of share in energy consumption (Economic Survey of Pakistan, 2019). The physical landscape—free rooftop space available in the household sector—further aligns to drive bottom VRE growth. Nevertheless, owing to lagging innovative organization, finance, and business models, the immense potential in this sector remains untapped.

Finally, there is flexibility-related impediments in the way making the country ill-prepared for any potential RE investments in the future. The Transmission and Distribution (T&D) sector in Pakistan lags way behind other energy sectors and is characterized not only by high losses, but also severe capacity constraints. Consequently, the Grid Interconnection approval is so time-consuming that it leads to unnecessary regulatory delays (IFC, 2016). And although, currently, Pakistan has not reached the point of VRE share which could create flexibility challenges, weak and insufficient grid infrastructure remains the most serious barrier confining renewable energy development in Pakistan. A substantial centralized renewable energy transition hence remains implausible owing to the absence of necessary resources and infrastructure including limitations surrounding renewable energy forecasting capacity isolated grid standing and the inefficient distribution network infrastructure characterized by outdated and obsolete technology.

Based on the above course of events, renewable energy policies, governance, and regulations have been quite fragmented in Pakistan. Further absence of innovative organization, business, and finance models, as well as grid-related constraints, has handicapped VRE growth. In addition, limited incentives and skepticism surrounding renewables further slow down the entire process. Collectively these forces lock the energy system into a carbon-dependent system, posing some of the thorniest challenges for any replacement or reversal making the transition highly resistant. Hence, the challenge is how best to complement policy tools with creating an environment for transition and timely overcoming fundamental barriers facing RE for effectively unleashing a renewable revolution in the country.

8.6 CONCLUSION

The energy landscape in Pakistan is undergoing a significant transformation. Overall, the generation capacity in the country has increased by around 40% alongside major alterations in the fuel mix. However, beyond creating a new generation and addressing the issue of a more diversified fuel mix, the disconnect between the dominance of fossil fuels in the power generation mix and the needed green development continues entrapping the country into a carbon lock-in. So, while much of the rest of the world transitions away from fossil fuels, Pakistan has further increased its use, further deepening its carbon trajectory which will remain locked in for several decades.

The CPEC infrastructure, especially the substantial investment in coal power plants, will have wide-scale environmental implications. Continual reliance on fossil fuels for economic growth and energy needs will come at the cost of a massive increase in GHG emissions which will exacerbate climate change, air pollution, and water scarcity challenges. The environmental concerns are particularly pressing in the light of Pakistan's commitments under the Paris Agreement, where it has committed to reducing up to 20% of its 2030 projected GHG emissions. Further, the availability of sufficient and affordable energy is critical to tackle poverty and improve living standards. Nonetheless, the share of fast-rising, expensive fossil fuel imports will have significant implications—creating a permanent headwind to economic growth and wider energy access.

The skewed development toward fossil fuels also reflects Pakistan's rhetoric on green development. Unstable policies and timid reforms for harnessing renewable energy have largely jeopardized the right environment in the sector. A balanced approach to the emerging challenges of a reliable and safe energy supply hence remains missing from the evolving energy landscape of Pakistan, with the transformation initiatives largely driven by the imperatives of energy accessibility. The existing trajectory in the power sector is not sustainable on all accounts, i.e., social, economic, and environmental. A sustainable energy sector trajectory would require the sector's circular debt (arrears across the power supply chain) to be addressed. In parallel, for the purposes of avoiding the potentially execrable outcomes associated with environmental degradation and energy insecurity, the energy transition has to be at the center of the agenda for sustainable development.

Bridging the gap between supply and demand in the country's energy sector requires a more balanced approach and strict environmental regulations while making developments. Overcoming the energy crisis should go hand-in-hand with a reduction in the dependence on fossil fuels and development in the RE sector. Finally, based on the intensity of environmental challenges in the country, if a transition does not occur soon, it may become too late to reverse the impacts of fossil dominated energy system.

The recently introduced Alternate Renewable Energy Policy (AREP) aims to take up the share of renewable energy to 30% of generation by 2030 compared to the current 4%. These changes at the policy level have sparked renewed hope to advance the clean energy transition and shelve the otherwise coal-based planned additions. The first cornerstone for energy transition has therefore been successfully laid with the introduction of the new AREP (AREP, 2019) and political support from policymakers. Variable Renewable Energy (VRE) has also recently become the cheapest source of power generation in Pakistan which further aligns with meeting the new target. Amid these dynamic forces, Pakistan could potentially reach greater heights of clean energy development. The real challenge—timely overcoming fundamental barriers facing VRE and creating a much needed enabling environment with regard to unleashing its development in the country—therefore remains. Having an enabling environment in place is not a trivial task where supportive policies, changes in existing regulations, technological improvements, a transformation of markets, financial resources, and new business models must be organized to successfully leverage the underlying transformational forces substantiating the transition. The challenge of advancing renewable energy hence cannot be achieved by a single government, institution, or sector; rather, it will require consolidated changes at every level of the energy value chain. A systemic approach to solving fundamental barriers and capturing opportunities is therefore imperative to foster the desired transformation.

To conclude, deepening energy sector reforms are required where the promotion of renewable technologies should constitute the cornerstone of the reform agenda. A sustainable long-term renewable energy-based model and a thoughtfully considered and prudent framework for accomplishing a well-balanced energy mix could mobilize a transition in no time.

REFERENCES

Small, Andrew. 2017. "First Movement: Pakistan and the Belt and Road Initiative," Asia Policy No. 24, 80–87.

AsianPower. 2019. "China Retains Crown as Most Attractive Renewable Energy Market."

ADB (Asian Development Bank). 2017a. "Climate Change Profile of Pakistan."

ADB (Asian Development Bank). 2017b. "PAK: 48078-002 MFF II Power Transmission Enhancement Program (PTEIP II)."

ADB (Asian Development Bank). 2019. "Pakistan: ADB's Support to Pakistan Energy Sector 2005–2017."

ADB (Asian Development Bank). 2018. "Green Finance in Pakistan: Barriers and Solutions."

Alternative and Renewable Energy Policy, (AREP), 2019, Government of Pakistan

Yukhananov, Anna and Valerie Volcovici. 2013. "World Bank to Limit Financing of Coal-Fired Plants," Reuters.

Bacon, Robert W. 2019. "Learning from Power Sector Reform: The Case of Pakistan." The World Bank.

Climate Change Division. 2013. "Framework for Implementation of Climate Change Policy (2014-2030)," Government of Pakistan, Islamabad.

CPEC Secretariat, Ministry of Planning, Development and Reform, "CPEC Projects," China-Pakistan Economic Corridor.

Dawn. 2019. "Alarm Mounts as Wind Power Turbines Halt after Govt Stops Purchases."

Eckstein, David, Vera Künzel and Laura Schäfer. 2018. "Global Climate Risk Index 2018," *Germanwatch*, Bonn.

Global Climate Risk Index 2020. 2019. *Bonn: Germanwatch* (2019).

HEI, I. (2017). State of global air 2017: a special report on global exposure to air pollution and its disease burden. 2017.

Eckstein, David, Vera Künzel, Laura Schäfer and Maik Winges. 2020. *"Global Climate Risk Index 2020," Germanwatch*, Bonn.

"Environmental performance index." *Yale University and Columbia University: New Haven, CT, USA* (2020).

Wendling, Z. A. et al. 2018. 2018 Environmental Performance Index. *New Haven*, CT: Yale Center for Environmental Law & Policy.

Faraz, Shibli. 2018. "Report of the Special Committee on Circular Debt." *Senate Secretariat.*

IFC. 2016. "A Solar Developer's Guide to Pakistan."

IFC. 2015. "Pakistan Off-Grid Lighting Consumer Perceptions Study," *World Bank Group.*

Illangovan, P., & Francisco, M. (2017). *Managing risks for sustained growth: Pakistan development update* (No. 121027, pp. 1-92). The World Bank.

IEEFA. 2018. "Pakistan's Power Future."

Health Effects Institute. 2017. "State of Global Air 2017: A Special Report on Global Exposure to Air Pollution and its Disease Burden."

IEA. 2017. *"World Energy Outlook 2017."* Paris: International Energy Agency.

IEA, IRENA, and WB UNSD. 2019. "WHO. Tracking SDG 7: The Energy Progress Report 2019."

Khalid, Zainab, and Muhammad Iftikhar-ul-Husnain. 2016. "Restructuring of WAPDA: A Reality or a Myth." *The Pakistan Development Review*, 349–359.

Muthukumara, Mani, Sushenjit Bandyopadhyay, Shun Chonabayashi, Anil Markandya, and Thomas Mosier. 2018. "South Asia's Hotspots: The Impact of Temperature and Precipitation Changes on Living Standards." *The World Bank.*

Murdock, Hannah E., Duncan Gibb, Thomas André, Fabiani Appavou, Adam Brown, Bärbel Epp, Bozhil Kondev et al. 2019. "Renewables 2019 Global Status Report."

Economic Survey of Pakistan (2002-2019). 2019. Ministry of Finance, Pakistan.

Economic Survey of Pakistan, (2019), Ministry of Finance, Pakistan

Economic Survey of Pakistan (2018), Ministry of Finance, Pakistan.

Economic Survey of Pakistan (2010), Ministry of Finance, Pakistan.

NEPRA. 2006-19. "State of Industry Report".

NEPRA. 2019, "State of Industry Report"

NEPRA. 2018a, "State of Industry Report"

NEPRA. 2018b. "Performance Evaluation Report of Distribution Companies."

NEPRA. 2017. State of Industry Report".

NEPRA. 2013. "State of Industry Report."

OGRA. 2018. "Petroleum Industry Report."

USAID. 2013. Planning Commission. "The Causes and Impacts of Power Sector Circular Debt in Pakistan."

Private Power and Infrastructure Board (PPIB). 2004. "Pakistan Coal Power Generation Potential."

Government of Pakistan. 2013. National Power Policy.

Raza, Babar. 2019. "Pakistan Can Produce Renewable Energy. So Why Do We Continue to Import Pricey Fossil Fuels?", www.dawn.com.

Khan, Rina Saeed. 2017. "Pakistan Races to Tap Virgin Coal Fields to Meet Energy Crunch," Reuters.

Sanchez-Triana, Ernesto, Santiago Enriquez, Javaid Afzal, Akiko Nakagawa, and Asif Shuja Khan. 2014. "Cleaning Pakistan's Air: Policy Options to Address the Cost of Outdoor Air Pollution." The World Bank.

Saleh, N. 2019. "Tracking Circular Debt", Institute of Policy Studies.

Express Tribune. 2018. "Power Generation Hits All-time High at 14,017 GWh." September 25.

USAID. 2012. "Green House Gas Emissions in Pakistan." Climate Links.

WRI. 2015. Ranking the world's most water-stressed countries in 2040. *World Resources Institute*, 26.

WRI (2016). Climate analysis indicators tool: WRI's climate data explorer. *World Resources Institute, Washington.*

World Bank-ESMAP (Energy Sector Management Assessment Program). 2014. "Renewable Energy Resource Mapping Pakistan."

Zhang, Fan. 2018. "In the Dark: How Much Do Power Sector Distortions Cost South Asia?". World Bank.

9 Energy, Environment and Sustainable Development Futures in Sri Lanka

Ajith de Alwis and Maksud Bekchanov

CONTENTS

9.1   Sri Lanka in Perspective 195
9.2   Energy Sector Dynamics 197
    9.2.1   Energy Supply 197
    9.2.2   Power Generation 199
    9.2.3   Energy Demand 200
    9.2.4   Energy Policy 201
    9.2.5   Energy Efficiency and Conservation 203
    9.2.6   New Renewable Energies 203
9.3   Environmental Management and Change 207
    9.3.1   Greenhouse Gas Emissions 207
    9.3.2   Deforestation and Biodiversity Loss 207
    9.3.3   Waste Management 208
    9.3.4   Climate Change Impact 210
    9.3.5   Environmental Policy 210
    9.3.6   'Green' Developments 211
9.4   Sustainable and Low-Carbon Development Futures 213
9.5   Conclusion 213
References 214

9.1   SRI LANKA IN PERSPECTIVE

Sri Lanka is an island country in South Asia and surrounded by the Indian Ocean along the coastlines extended over 1,340 km. The island is located about 80 km to the Southeast of the Indian subcontinent and separated from mainland India by the Palk Strait (Figure 9.1). The country's land area covers 65,610 km^2 and its maritime waters reach 489,000 km^2 (Munasinghe 2019).

Despite its small size, Sri Lankan geography comprises a wide range of agro-ecological landscapes and is characterized with very rich bio-diversity. Ecosystems range from mountainous rainforests to semi-arid plains, rivers, lakes, coastal areas, and marine environment. The mountainous zone elevated up to 2,524 meters in the center of the island is a main water source for major rivers spread across the country. Coral reefs and sea grass beds span along coastal areas and adjacent waters.

FIGURE 9.1   Geographic position of Sri Lanka. (Source: Australian National University 2019.)

The country is rich in minerals such as graphite, ilmenite, silica, feldspar, kaolin, and thorium. Recent explorations confirmed also the availability of fossil fuel resources in the Gulf of Mannar (CEB 2018; Premarathne et al. 2013).

Changes between a wet season with heavy monsoon rainfall and a dry season with less precipitation characterize the climate in the country. Precipitation is heterogeneous across the island reaching 5,000 mm in the central zone and less than 1000 mm in the northwest. The mean annual temperature is also variable across the country varying from 26 to 28°C in coastal regions and ranging from 5 to 19°C in the center.

Legislative and administrative structure in Sri Lanka is based on a Constitution. The country comprises nine administrative provinces. Colombo in the Western province is the capital city with rapidly growing infrastructure and a vibrant economy. The recent addition of 269 ha of reclaimed land—the Port City—enhances the economic capacity of the capital further. The strategic location of the island at the main routes of global trade opens new ways of reaping benefits from the integration into the global economy.

Sri Lanka is a home for 21.9 million people (as of 2019) and annual gross domestic product (GDP) per capita raised from US$897 in 2000 to US$4102.5 in 2018. After resolving the long-lasting and nationwide civil conflict in 2009, the country's economy attracted considerable foreign investments and has been steadily growing. As a result, the country was able to raise its status to an upper middle-income country in a relatively short time period. Human development indicators have shown excellent improvements in the country. The prevalence of poverty (percentage of population earning less than US$1.99 per day per person) decreased to less than 2%, and it is the country with the lowest poverty rate in the region. Nevertheless, the number of the people living under life conditions close to the poverty line is considerable and income inequality is high. Rapid economic development and urbanization also raised energy intensity (more than two times between 2000 and 2015) and escalated environmental pollution (Munasinghe 2019). As the analysis of energy security across Asia for the period of 1990–2014 indicated, Sri Lanka is one of the countries with gradually decreasing energy security (Le Thai-Ha et al. 2019). These issues should gain proper attention for designing current and future government policies aimed at sustainable and inclusive development.

9.2 ENERGY SECTOR DYNAMICS

9.2.1 ENERGY SUPPLY

To meet the demands of rapid economic growth, energy supply increased in Sri Lanka from 413 PJ in 2005 to 528 PJ in 2017 (Figure 9.2). The primary sources of the energy changed from traditional biomass to fossil-based fuels over time. If biomass share in overall energy supply decreased from 60% to 40% from 1990 to 2017, the share of the fossil-based fuels (petroleum and coal) rose from 30% to 50% in the same period. Renewable energies (solar, wind, hydropower) contribute to relatively small, yet considerable, share of total energy supply. Regulating the expanding energy sector, the government established the Public Utilities Commission of Sri Lanka (PUCSL) in 2002.

Since the country has currently no exploitable fossil-fuel reservoirs, fossil-based energy is imported from abroad either in raw (crude) or processed form. Sapugaskanda Refinery is the only refinery that converts the imported crude oil into refined products and produces 2.3 million tons of outputs annually (SLSEA 2018).

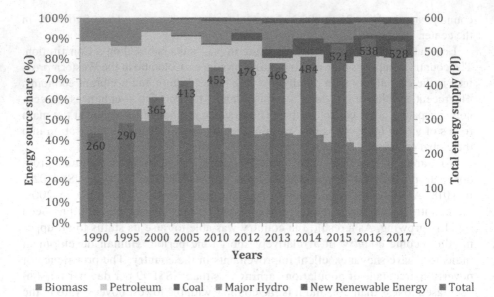

FIGURE 9.2   Dynamics of energy supply and sources. (Source: SLSEA 2018.)

As the refinery did not go through proper modernization except some debottleneck-ing since its commissioning in 1969, the share of refined products currently domi-nates in total energy imports (Table 9.1). The balance of payment in the country is adversely impacted by the imports of the energy commodities and high share of refined products in the imports.

TABLE 9.1

Imports of Energy Commodities (2017–2019)

Quantity Imported	Unit (tons)	2017	2018	2019
Crude Oil	1000	1,591	1,674	1,842
Refined Products	1000	4,895	4,959	4,740
Coal	1000	2,530	2,167	2,390
Liquid Petroleum Gas (LPG)	1000	397	413	430
Value of Imports (Cost, Insurance and Freight)				
Crude Oil	Rs. M	107,397	160,024	173,547
	US$ M	704	978	971
Refined Products	Rs. M	375,374	475,521	483,462
	US$ M	2,462	2937	2,706
Coal	Rs. M	39,699	38,750	38,719
	US$ M	261	237	215
LPG	Rs. M	35,505	42,162	43,156
	US$ M	233	266	241

Source: Ceylon Chamber of Commerce – Economic Intelligence Unit,.

9.2.2 POWER GENERATION

The national electricity grid covers the entire area of the mainland island. The grid connectivity level in Sri Lanka is 98.7% (Athukorale et al. 2019). Therefore, in terms of energy access, the country's performance is one of the best among South Asian countries. The transmission and distribution (T&D) loss is about 8.8%. While potential for further development of the electricity grid is limited in the mainland, there is a large potential for off-grid electricity generation in small islands belonging to the country.

The Sri Lankan power system has a total installed capacity of approximately 4,093 MW with an annual output of 15,000 GWh (as of 2017; Figure 9.3). Main suppliers to the electricity grid are thermal (oil- and coal-based) and hydropower plants. Almost 50% of the electricity generation has come from carbon-intensive (primary) energy sources (as of 2017). The Ceylon Board of Electricity (CEB) is in charge of managing the national power system and owns a majority of the large thermal and hydropower plants. Seven oil-based thermal plants under the coordination of the CEB generate about 2,600 GWh of electricity annually, while the coal-fired power plant (3x300 MWe units) generate over 5,100 GWh (SLSEA 2018).

Rivers from the hills of the central zone of the island are essential sources of hydropower. Major hydropower complexes are Laxapana (on the Kelani River) and Mahaweli (on the Mahaweli River), each consisting of numerous power generation plants (SLSEA 2018). These complexes, together with the plants in other river basins, generate about 3,000 GWh of hydropower annually. The development of the large hydropower projects has reached almost its full potential in the country. In contrast, newly emerged renewable energy sector has been steadily expanding as climate and geography of the country create favorable conditions for solar, and wind power generation.

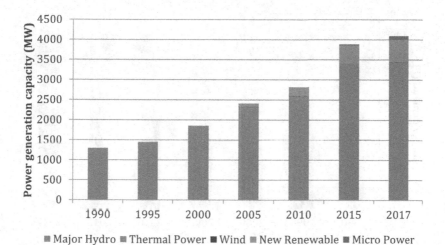

■ Major Hydro ■ Thermal Power ■ Wind ■ New Renewable ■ Micro Power

FIGURE 9.3 Power generation capacity development (1990–2017). (Source: Based on SLSEA 2018.)

Due to the lack of sufficient capacities of the CEB to meet the recently grow-ing energy demands, independent power producers (IPPs) and small power produc-ers (SPPs) are allowed to produce electricity and supply to the grid. While most of the large IPPs are oil-based thermal power plants, SPPs manage micro hydropower plants, a solar power plant, and numerous biomass-based plants. IPPs and SPPs can enter into contractual terms with the CEB to sell their energy outputs to supply to the national grid.

9.2.3 ENERGY DEMAND

The residential sector (households, commercials and others) was a key user of the gen-erated energy in Sri Lanka accounting for over half of total energy demand until 2005 (Figure 9.4). Energy demand substantially increased by the transportation and indus-trial sectors between 2005 and 2017, reaching up to 36% and 24% of the total energy demand, respectively. Consequently the total energy demand in Sri Lanka has increased at the rate of 1.9% per annum changing from 336.8 PJ in 2005 to 423.8 PJ by 2017.

The transport sector fully relies on fossil-based fuels (as of 2017). Only 16% of the industrial demand is met by electricity. A total of 60% of biomass energy is primar-ily consumed as fuel wood for cooking, and the remaining goes to industry, mainly for generating steam power (for instance, for laundry services in hotels) or biogas (for cooking in staff kitchens). At present, although 42% of the households have switched to liquid petroleum gas (LPG) for domestic cooking, biofuel sources account for nearly three-fourth of household energy demand.

The average per capita electricity consumptions were 603 and 626 kWh per per-son in 2016 and 2017, respectively. The electricity demand has been growing at an annual rate of around 5–6% during the past 20 years. Improved incomes increase the purchases of electricity appliances, consequently leading to higher electricity

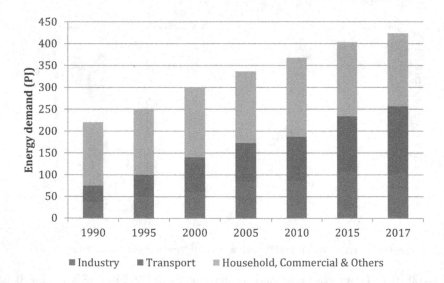

FIGURE 9.4 Dynamics of energy demand (1990–2017). (Source: Based on SLSEA (2018).

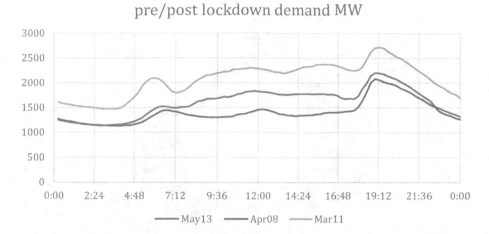

FIGURE 9.5 Daily electricity demand curves in March, April, and May. (Source: Wickramasinghe H., Personal Communication, 12/05/2020.)

consumption. The government also charges much lower customs fees to the imports of hybrid cars that can run using either petrol or electricity, making them more marketable compared to the petrol-based cars. Thus, electricity consumption is expected to increase in the transportation sector over time.

Energy consumption patterns also vary by months and hours of the day. The typical energy demand profile of the country is presented in Figure 9.5. One major observation on the demand curve is the peak demand is between 18 and 20 pm. It is not surprising that electricity demand for lighting, cooking dinner, and watching TV simultaneously increases during these hours. Daily electricity demand also considerably varies across the months. Increased temperature in certain months may require cooling and more intense uses of the air conditioners.

9.2.4 ENERGY POLICY

Governmental energy policies address the energy security concerns and determine the change in energy use patterns. A National Energy Policy has been shaped over a long time in Sri Lanka. Initially, energy policies focused on careful energy uses and national energy self-sufficiency over a long time aiming at an "energy conscious, secure and energy non-dependent nation". Present policies concentrate on the energy system transition with increased uses of renewable energies and improved energy efficiency (UNDP 2017, UN-ESCAP 2019). These targets determine the course of the energy plans and related measures documented in the 2019 National Energy Policy and Strategies for Sri Lanka, National Policy on Transport in Sri Lanka, National Energy Management Plan (NEMP), Long Term Generation Expansion Planning Studies (LTGEPS), and Energy Sector Development Plan (ESDP).

ESDP considers increasing the share of renewable energies in electricity supply to 60% by 2020 and gradually achieving the level with 100% renewables by 2030. This ambitious goal of transitioning into the state of "100% electricity from renewable

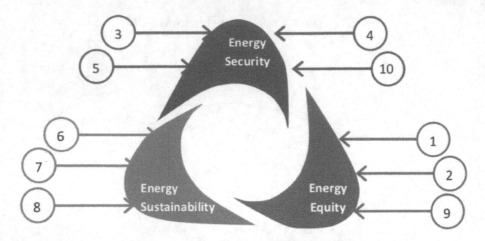

FIGURE 9.6   National energy development framework.

energy sources" is quite environmentally friendly, but it may come at a very high cost, requiring investments amounted to US$50 billion (UNDP 2017).

The 2019 National Energy Policy and Strategy recently published documents with the goals and necessary tasks for ensuring sustainable development of the energy sector in Sri Lanka. Ten pillars are pointed out to ensure security, equity and sustainability of the energy services (Figure 9.6):

1. Assuring energy security
2. Providing access to energy services
3. Optimizing costs of supplying energy to the national grid and providing energy services
4. Improving energy efficiency and promoting energy conservation
5. Enhancing self-reliance
6. Caring for the environment
7. Enhancing the share of renewable energy
8. Strengthening the good governance in the energy sector
9. Securing land for future energy infrastructure
10. Providing opportunities for innovation and entrepreneurship

In addition to supporting sustainable development of the energy sector, government policies are currently allowing for wider involvement of the private sector in energy developments. Until the recent past, the Sri Lankan energy sector has been primarily a state monopoly, and the role of the private companies in the sector was limited. Recently, crude oil and refined fuels have been opened up for the private sector providing entitlement to the Indian Oil Corporation, which is known in the country as LIOC or Lanka Indian Oil Corporation. LPG imports and distribution across the country are also entrusted to a private sector (Litro Gas Lanka Ltd, LAUGFS gas PLC, and LIOC Plc). As already discussed above, IPPs and SPPs have been also active in the renewable energy sector.

9.2.5 ENERGY EFFICIENCY AND CONSERVATION

Ensuring energy efficiency and conservation gained attention in governmental policies following the 1973 Global Energy Crisis and has not lost its importance in the national energy development agenda since then. Promoting the energy efficiency and conservation measures and the uptake of the renewable energy sources at the national level are entrusted to the Sri Lankan Sustainable Energy Authority (SLSEA).

For reducing energy demand through the implementation of energy conservation measures, the SLSEA pursues the following strategies:

- Improving energy efficiency and implementing the conservation measures wherever relevant along the entire energy supply chains;
- Formulating policies and strategies to encourage transition of the national energy system by reducing the reliance on fossil-fuel uses and scaling up sustainable energy options;
- Enabling environment for sustainable energy investments in the country;
- Introducing and promoting new sustainable energy technologies; and
- Raising public awareness to promote healthy lifestyle and sustainable living environment.

The SLSEA has significantly contributed to design and implementing the measures for improving energy efficiency of electric appliances, technologies and processes in all economic (industrial) and residential (household, commercial and others) sectors. It has identified 11 thrust areas, which can greatly enhance energy savings at the national level (10% reduction in energy demand): Efficient Lighting, Fans, Refrigerators, Air Conditioning, Pumps, Motors, Eliminating Incandescent Lamps, Green Buildings, Energy & Building Management Systems, Smart Homes, and Power Factor (Generation Efficiency) Improvement. The SLSEA is currently working on the certification of energy efficiency labeling for most of these appliances in cooperation with Sri Lanka Standards Institute. Energy conservation in commercial, industrial and domestic sectors is enforced through introducing ISO 50001–Energy Management System (EnMS). Moreover, energy efficiency enhancements are supported through a legislative mechanism of placing energy managers in high- energy consuming installations. SLSEA has also promoted energy efficiency and conservation behavior through study programs in the primary and secondary education systems.

9.2.6 NEW RENEWABLE ENERGIES

Sri Lanka as a tropical country has large potential for reaping benefits from the utilization of the renewable natural resources such as hydropower, solar power, wind power, and biomass (Figure 9.7). Maps developed by SLSEA also confirm considerable power generation potential of different renewable energy sources across Sri Lanka. As the country is blessed with several rivers and abundant water supply, hydropower meets a considerable share of the national electricity demand. Vagaries of weather and seasonal droughts in recent times, however, reduced the reliability of the hydropower option.

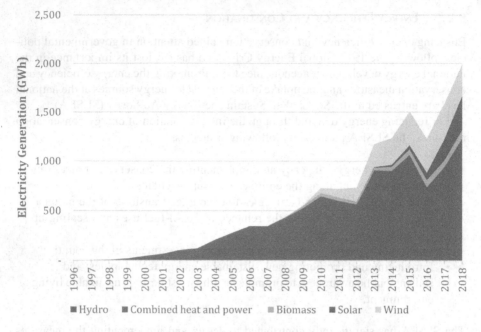

FIGURE 9.7 Dynamics and structure of the new renewable energy sector. (Source: SLSEA 2018.)

The country has enormous biomass-based energy sources. Biomass-based energy is widely used for household cooking and for thermal power generation in industry. In addition to firewood, bagasse, coconut shells, and organic waste are essential sources for energy. Although with less success so far, bio-fuel crops such as *Gliricidia sepium* were also promoted in Sri Lanka aiming at development of the national bio-energy sector.

Biogas generation technology has currently lower power generation capacity compared to the solar or wind power generation options (Figure 9.7). Yet, testing the biogas technology potentials to improve national energy security has a longer history going back to the 1970s. When the United Nations (UN) convened a meeting in Colombo at that time after the energy crisis of 1973, biogas was identified as the renewable energy source to scale up as indicated in the Colombo Declaration. Thus, Sri Lanka was chosen as a model country by UNEP to implement the biogas technology. Biogas technology has several advantages including supplementary energy for household cooking, possibility to use as transport fuel, and recycling organic waste and biomass (de Alwis 2012). Hot temperature in the Tropics is also favorable for the effective functioning of the anaerobic bacteria, increasing the efficiency of biogas generation from biomass (household organic waste, wastewater, crop residues, and livestock waste). As estimated, biogas from organic waste can supply energy equivalent to US$450 to US$630 million in economic terms and can substitute 20 to 30% of the biomass-based household energy demand in Sri Lanka (Bekchanov et al. 2019). However, due to several financial, technical, legal, and socio-cultural constraints, the adoption of the technology is quite slow in the country.

In the early days of the implementation of the off-grid solar energy technology, Sri Lanka was considered as a model country for successful implementation of solar technology (Miller 2011). Year-round supply of solar radiation in the country provides invaluable opportunities for wider adoption of solar energy technologies. With the reduced costs of the solar panels over time, solar technologies became more financially feasible. The *Surya Bala Sangramaya* (meaning "Solar Energy Battle" in the local language) program introduced by government in 2016 considers promoting solar power installations with the overall capacity of 200 MW by 2020, and 1000 MW by 2025 (Jayaweera et al. 2018). Progressive charges to electricity use are applied to reduce wasteful electricity uses and encourage the households with high energy demand to implement renewable energy technologies. To increase the profitability and thus boost wider adoption of the rooftop solar installations, the CEB introduced power purchase schemes that allowed selling the surplus power produced by the adopters. This policy is encouraging an increasing number of the households and organizations to choose installing rooftop solar panels in their premises (Figure 9.8).

The *Surya Bala Sangramaya* program also promotes constructions of solar power parks (SPPs). The first commercial SPP of the country, with the capacity of 500 kW, was established in Hambantota in 2016, with the aid of the Korean government. Total capacity of the commercial solar power plants gradually became 51 MW by the end of 2018. When considering the rooftop solar installations, overall solar power generation capacity reached to 170 MW in the country. By the end of 2018, development of 37 small-scale SPPs with the individual capacity of 1 MW and using private investments had been approved by authorities. Another 90 small-scale SPPs

FIGURE 9.8   Rooftop solar power installations. (Source: Photo by the authors.)

with the individual capacity of up to 1 MW have been under evaluation to launch during 2019.

Wind power generation is also increasing gradually in Sri Lanka. Three major regions in Sri Lanka are identified as having good-to-excellent potential for generating wind power: the Northwestern coastal region from Kalpitiya Peninsula to Mannar Island, the Jaffna Peninsula, and Central highlands in the interior of the country (NREL 2003). Recent estimates indicate that the country owns nearly 5,000 km^2 of windy areas with good-to-excellent wind power generation potential. An initial wind power plant with 3 MW capacity was established at Hambantota with the aid of a UNEP grant. The first commercial wind power plants were established in 2010 and the total capacity of wind power reached 128 MW at the end of 2018. A new wind farm with the capacity of 100 MW is now at the implementation stage in Mannar Island.

Ocean Thermal Energy Conversion (OTEC) is another option which should be investigated further. It is known that Sri Lanka is one of the best locations in the world for the development of the OTEC plants. Particularly, Trincomalee has sites with excellent depth closer to the shore to implement OTEC projects. Interest in promoting OTEC and other unconventional renewable energy options is currently on the rise. To facilitate novel renewable energy concepts, CEB has recently called for Expression of Interest from prospective private developers working in the fields of geothermal energy conversion, compressed air based power generation, Ocean Thermal Energy Conversion (OTEC), ocean energy (wave) conversion, biogas power generation and other storage applications, such as grid scale battery storages for energy shifting and hybrid systems that integrate wind and solar photovoltaic (PV).

Despite numerous environmental benefits of the new renewable energy options, their high investment cost requirement is a key hurdle for scaling up the adoptions. High interest rates of the bank loans and limitedness of the subsidies reduce the financial feasibility of the renewable energy projects. Financing renewable energy projects consequently can be quite risky at competitive interest rates and less attractive to the private sector representatives (World Bank and International Finance Corporation 2019).

Most of the current renewable energy developments in the country have been realized as a result of international donor funding. With the current change in economic status of the country being classified as an upper middle-income country, such funding schemes will no longer be available. Yet, there are also some business companies such as hotels which were able to create an "eco-friendly" brand for their services through adopting renewable energy technologies and energy efficiency improvement measures. This definitely improves their incomes as environmentally conscious tourists choose their services to support the "green growth" programs.

Comparisons with other non-renewable alternatives indicate that coal-fired power plants generate energy at much lower costs. Despite this fact, environmental externalities related to the coal-firing should not be forgotten. As the carbon-intensive power generation options are being gradually phased out in the developed countries, coal-fired energy businesses may seek a refuge in developing countries. Therefore, environmental externalities related to air, soil and water pollution should not be avoided in assessing the feasibility of the power generation projects.

9.3   ENVIRONMENTAL MANAGEMENT AND CHANGE

Living mindfully and in harmony with nature was an essential aspect of the culture in Sri Lanka over centuries. While inventing various effective ways of utilizing natural resources for sustainable livelihoods, environmental safeguarding and careful approach to nature were not forgotten. Early civilizations in the country developed advanced hydraulic systems by establishing a network of small tanks and water reservoirs, which maintained flourished irrigation and ecosystems. Human activities related to irrigation development and settlements are well designed considering the fauna and flora of the site and did not damage the naturally beautiful ecosystem landscape of the country. Thus, the country deservingly gained the name "Serendip" which means "good fortune". However, development programs over the last century, partially inherited from the country's colonial past, greatly modified the approach toward the environment. At present, key environmental problems such as air and water pollution, climate change-related drought and flooding, degradation of forests and coastal ecosystems, and urban waste disposal threaten sustainable livelihoods and require immediate action to prevent larger socio-economic damages.

9.3.1   GREENHOUSE GAS EMISSIONS

Being a key driver of the economy, the energy sector has a considerable environmental impact both at local and global levels. As the country largely imports fossil fuel-based energy commodities, it contributes to the carbon gas emissions in the regions where mining and refining took place. Large amounts of GHG emissions occur during the use of the fossil fuel-based energy in transportation and industrial sectors of the country. As estimated, overall GHG emission in the country is 16.7 million tons per annum (as of 2014), or 0.6 tons per capita, of which 41% come from the energy sector.

Within the energy sector, coal-fired power generation became the biggest contributor to the total GHG emission pool (Figure 9.9). Due to the large contribution of hydropower for the national energy mix, the GHG emission rate was previously low in Sri Lanka, and the country had sound performance according to the international environmental ratings. Following the commissioning of the coal-fired power station with the capacity of 900 MW in 2010, GHG emissions considerably increased. In result, GHG emission intensity of the national power system also increased from 316 to 585 gCO_2 per kWh between 2010 and 2019.

9.3.2   DEFORESTATION AND BIODIVERSITY LOSS

Deforestation and land use change in the country also considerably contribute to the increasing GHG emissions (14% of the total) and have adverse effects on biodiversity. Though the deforestation rate is low at present, forest cover was substantially (30–40% loss) depleted due to hydropower developments, expansion of the residential settlements, and clearing forests for irrigation and agriculture during the last

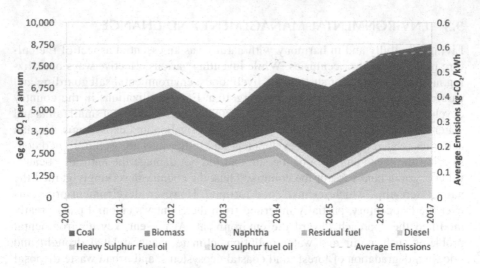

FIGURE 9.9   GHG emissions from the energy subsectors. (Source: SLSEA, 2018.)

century. Currently, in addition to encroachment for residencies and for planting tea or other field crops, growing demand of timber and sawnwood by households and industries are main factors for deforestation.

Decline of rainforests in turn change the local climate increasing the risks of water availability anomalies. Deforestation also intensifies soil erosion and land sliding occurrences subsequently increasing sedimentation of the reservoirs and threatening agricultural and hydropower production.

Encroachment for living spaces, especially when the living areas intersect the seasonal elephant travel routes, leads to the conflicts between humans and wildlife. At present, the human–elephant conflicts take the lives of 200 elephants and 80 people annually. Nevertheless, recognizing the importance of the wildlife conservation, some projects were also undertaken to support the wildlife. An artificial lake in a natural reserve–Minneriya, for example, is a unique elephant gathering site allowing the animals to survive during the drought period (Figure 9.10).

Deforestation and degradation of the natural ecosystem landscapes are also a key threat to biodiversity loss. Given the rich biodiversity and natural endowments, Sri Lanka is considered as one of 34 biodiversity hotspots across the world. The country has 48 distinct agro-ecological regions and hosts a great array of endemic species. From a conservation standpoint, the number is significant and their extinction would represent an irreplaceable loss. The recent Red List by the International Union for Conservation of Nature (IUCN) identified 66 critically endangered and 102 endangered animal species (elephant and leopard are two flagship species) and 156 critically endangered plant species.

9.3.3   Waste Management

GHG emissions from open dumping of municipal and household waste are considerable accounting for 26% of total emissions in Sri Lanka (UNDP 2017). Total solid

FIGURE 9.10 Annual elephant gathering in Minneriya.

waste generation is about 9810 tons per day in the country (collected 3767 t/d). Per person waste generation in urban settings is about 0.72 kg/d with the general average at 0.5 kg/d. Although average waste generation rate per capita at the national level is low due to the small share of manufacturing sector in the national economy, concentration of the waste per capita is quite high in urban areas, especially in the capital city and surroundings.

Poor management and open disposal of waste are serious threats to the environment and livelihoods. In 2017, the Meethotamulla Open Dump ('garbage mountain') in the surroundings of Colombo collapsed killing more than 33 people and is entered in history as one of the worst waste management failures in the world (Jayaweera et al. 2019). In addition to life-threatening effects and air pollution impacts, open waste dumping into the areas adjacent to lakes and wetlands also pollutes fresh waters (Hettiarachchi et al. 2014). As wastewater is discharged into the Indian Ocean from all coastal areas and from Colombo City without adequate treatment, it has adverse impacts on marine ecosystems.

Wider implementation of circular economy technologies can effectively cope with environmental externalities related to the open dumping. Since the share of organic content in municipal and household waste is high in Sri Lanka, recycling organic waste through anaerobic digestion or composting processes allows for minimizing environmental impacts. Organic waste recycling can supply valuable energy and nutrients (organic fertilizer) to the domestic economy, enhancing energy and food security (Bekchanov and Mirzabaev 2018; Bekchanov et al. 2019).

All other solid waste streams such as e-waste, healthcare waste and hazardous waste go to a separately managed system (Mallawarachchi et al. 2012). Due to the presence of a multinational waste processing company, Insee Ecocycle, with a large capacity, considerable share of the total hazardous waste streams are currently being

TABLE 9.2

Climate Risk Assessment in Sri Lanka

Climate Risk Assessment Parameter	Value
Climate Risk Index (2017)	2
Population (M)	21.4
Emissions (tons per capita; 2014)	1.37
Climate Vulnerability Level	severe
Mortality attributable to air pollution (per 10,000; 2015)	125.4
Deaths attributable to climate change (2015)	627
Child deaths attributable to climate change	6.06
Economic loss (% GDP)	1.14

co-processed and or co-fired (Gunarathne et al. 2019). Regulations are enforced to prevent harmful effects of these types of waste.

9.3.4 CLIMATE CHANGE IMPACT

Although GHG emission of Sri Lanka is quite small compared to those in the developed countries, the country faces serious challenges related to climate change impact. According to the most recent Global Climate Risk assessment, Sri Lanka is the 2nd most vulnerable country to the potential climate change impacts (Table 9.2). Despite questionable accuracy of the ranking, the country is prone to many climate-related natural calamities. Increased frequencies and magnitudes of the floods and droughts cause large economic losses (1–2% of the national economy) in Sri Lanka. Consequent loses of human lives are also severe. Recent flooding of the Kelani River in May 2016, for instance, severely affected 24 districts and caused landslides, which took 93 lives and damaged 58,000 houses. Flooding in May 2017 affected 15 districts, taking 212 lives and fully destroying 2,313 houses.

9.3.5 ENVIRONMENTAL POLICY

For preventing the adverse impacts of economic developments and environmental changes, the Sri Lankan government is seriously committed to maintaining environmental safeguarding. Sri Lanka also practices the pre-project environmental impact assessment for the industrial projects and has also a history of carrying out strategic environmental impact assessment. Different institutional reforms, sectoral policies and strategies were introduced to manage various environmental issues over years. Particularly, regulations on controlling air pollutant emissions have been priority in policy agenda. Climate Change Secretariat was established to record GHG emissions, implement a Carbon Development Mechanism (CDM) and manage the Sri Lankan Carbon Fund. As a signatory to the Kyoto and Paris accords, three national communications on GHG inventories have been completed.

For maintaining environmental sustainability, several regulations and policies have been adopted, such as:

- Clean Air 2000 Strategy and Action Plan (1992)
- National Policy on Protection and Conservation of Water Sources, Their Catchment and Reservations In Sri Lanka (2014)
- National Policy for Rural Water Supply and Sanitation Sector
- National Policy and Strategy on Cleaner Production for Agriculture Sector
- National Policy and Strategy on Cleaner Production for Tourism Sector
- National Land Use Policy
- National Watershed Management Policy
- National Policy on Wetlands
- National Forestry Policy
- National Biodiversity Strategy and Action Plan (2016–2022)
- Siting of High Polluting Industries
- National Waste Management Policy
- National Policy for Disaster Management
- The National Climate Change Policy of Sri Lanka
- The National Climate Change Adaptation Strategy for Sri Lanka

In addition to the governmental policies, active support by the citizens is also important for the transition to a sustainable and low-carbon economy (Gunarathne et al. 2020). Citizens can greatly impact the economic and technology change tendencies as consumers through adjusting their consumption patterns and behaviors. As a developing economy but a highly literate society, Sri Lanka has great potential to empower the environmentally conscious citizens and awaken the motivation for sustainable transformation.

9.3.6 'GREEN' DEVELOPMENTS

At present, various initiatives are taking place to support transitions toward green and low-carbon economy. The RAMSAR Convention on Wetlands that aims at protecting water bodies entered into force in Sri Lanka in October 1990, and six ecosystem areas were accredited wetland protection sites. In October 2018, the Commercial Capital Colombo was also accredited a RAMSAR wetland site. At present, many development projects are in progress aiming at maintaining a harmony between urban lifestyle and wetland ecosystems.

To maintain balanced interactions between human and nature systems, green constructions and practices are currently gaining traction in Sri Lanka. The number of green buildings has been increasing in textile and apparel industry, banking sector, and even in research and education sectors (Goger 2013). The country hosts the world's first platinum-rated apparel sector factories (one brand new and another retrofitted). The tallest vertical garden building was just recently constructed in Rajagiriya (Figure 9.11). Hotel Kandalama in Dambulla was built based on green architecture concepts and the first construction certified by LEEDS outside of the United States (Figure 9.12). The Green Building Council of Sri Lanka (GBCSL) is in charge of providing incentives in greening buildings and services, and running the country's green rating system.

FIGURE 9.11 Clearpoint residencies; vertical garden building in Rajagiriya, Colombo. (Source: GBCSL, 2019)

FIGURE 9.12 Green architecture; Hotel Kandalama in Dambulla, Central Province.

9.4 SUSTAINABLE AND LOW-CARBON DEVELOPMENT FUTURES

Sri Lanka is a signatory to various international conventions on environmental protection and actively supports sustainable development measures across the economic sectors. The Sri Lankan government endorsed the global "Agenda 21", the "Rio+20 outcomes" supporting the "10-year framework of programs on National Sustainable Consumption and Production Policy (2020)", and the "2030 Agenda for Sustainable Development". Sri Lanka was the first country in the world to have a Ministry entrusted with the promotion of sustainable development (in 2015). A bill of sustainable development was also presented to the Parliament later in 2017. The bill had provisions for the development and implementation of a national policy and strategy on sustainable development, and the establishment of a sustainable development council, at the highest level under the aegis of the president (SLSDB 2017). The council is assigned a duty of formulating the national sustainable development strategy. Every government agency is required to prepare a sustainable development task in accordance with the national policy and report the progress.

Regarding the energy sector, national sustainable development policies consider transition to low-carbon renewable energy production and uses (MoMD&E GoSL 2016). Nationally determined contributions (NDCs) to mitigate GHG emissions were agreed upon. In accordance with the nationally appropriate mitigation actions (NAMAs), 52 mitigation options were identified with a total potential of reducing 75 $MtCO_2e$ during 2020–2030 (Sugathapala et al. 2020). The energy sector accounts for 53% of the total GHG mitigation potential. In addition to energy-saving interventions, energy development projections consider enormous expansion of solar and wind power technologies over time to reduce the GHG emissions and maintain sustainable development (UNDP 2017). Some studies proposed also supporting renewable energy developments and energy security through establishing South Asian electricity grid, cross-border energy trade, and trans-regional investments (Shah et al. 2019). Yet, these assessments are at an exploratory stage currently and need further investigations.

Transport, industry, forests and waste management sectors also have considerable potential to mitigate GHG emissions. Particularly, in the transport sector, transitioning to environmentally sustainable public transport system, improving the roads and traffic systems by reducing traffic congestions can largely reduce energy demand and related GHG emissions. Replacing fossil-based fuel with biofuel can also greatly contribute to GHG emission mitigations. Some plans are already there, as the country expects transforming 25–40% of public transport into "green" fuel uses by 2030 (UNDP 2017). For enhancing reduced air pollutant emissions, the government also charges reduced customs fees to the imports of electric and hybrid cars.

9.5 CONCLUSION

Economic progress increases energy requirements and environmental footprints, consequently demanding actions to reduce environmental externalities in Sri Lanka. According to the statistical reports, energy use increased from 413 PJ to 528 PJ between 2005 and 2017. To meet energy demand, coal-fired power generation also expanded, almost doubling the GHG emissions. Wider implementation of the renewable energies including solar, wind power and anaerobic digestion of organic waste has large potential to enhance energy security while having minimal environmental impacts. At present, generating 2,000 MWh of electricity, the new renewables sector (solar, wind, biomass, small hydro) contributes to 10–15% of total electricity supply. Supporting green infrastructure is also an option for reducing energy demands and ideal to maintain harmony of the economy and lifestyles with nature, considering the country's environmental conditions. A holistic perspective while considering the interdependencies of the energy, environment and economic sectors allows for synergetic solutions, which are concealed or perceived as a problem when managing the sectors in isolation. The government has been always supportive to promote sustainable development measures by adopting relevant national regulations and being a signatory of various international conventions on ecosystem diversity and environmental protection. Due to high investment costs and inconsistencies in the governance, however, the implementation rates of the sustainable technologies are still slow. Nevertheless, if the steps with a long-term perspective are taken further with a strong commitment by the government and involving all citizens, Sri Lanka has enormous potential to be a model nation of sustainable development.

REFERENCES

Athukorale W, Wilson C, Managi S and Karunarathne M (2019), Household demand for electricity: The role of market distortions and prices in competition policy, Energy Policy, 134, 110932.

Australian National University (CartoGIS Services, College of Asia and the Pacific) (2019), Map - The island nation of Sri Lanka (formerly Ceylon) showing province borders and major cities.

Bekchanov M and Mirzabaev A (2018), Circular economy of composting in Sri Lanka: Opportunities and challenges for reducing waste related pollution and improving soil health, Journal of Cleaner Production, 202, 1107–1119.

Bekchanov M, Mondal A H Md, de Alwis A and Mirzabaev A (2019), Why adoption is slow despite promising potential of biogas technology for improving energy security and mitigating climate change in Sri Lanka? Renewable and Sustainable Energy Review, 105, 378–390.

Ceylon Electricity Board (CEB) (2018), Long Term Generation Expansion Plan 2018-2037, Colombo, Sri Lanka.

de Alwis AAP (2012), A tool for sustainability: A case for biogas in Sri Lanka, Journal of Tropical Forestry and Environment, 2, 1, 1–9.

National Sustainable Consumption and Production Policy (Draft, 2020), http://mmde.gov.lk/web/images/pdf/2018/EPE/scp policy english draft.pdf (accessed on 22/05/2020).

Goger A (2013), The making of a business case for environmental upgrading, Sri Lanka eco-factories, Geoforum, 47, 73–83.

The Green Building Council of Sri Lanka (GBCSL) (2019). Photo of vertical garden building in Rajagiriya (Prof. Ajith de Alwis was among the members of the expert panel for the national green rating system)

Gunarathne A D N, Kaluarachchilage P K H and Rajasooriya S M (2020), Low carbon consumer behavior in climate vulnerable developing countries: A case study of Sri Lanka, resources, Conservation and Recycling 154, 104592.

Gunarathne N, de Alwis A and Alhakoon Y (2019), Challenges facing sustainable urban mining in the e-waste recycling industry in Sri Lanka? Journal of Cleaner Production, 251, 119641.

Hettiarachchi M, Athukorale K, Wijekoon S and de Alwis A (2014), Urban wetlands and disaster resilience of Colombo Sri Lanka, Int. Journal of Disaster Resilience in the Built Environment, 5, 1, 79–80.

Jayaweera M, Gunawardana B, Gunawardana M, Karunawardena A, Dias V, Premasiri S, Dissanayake J, Manatunge J, Wijeratne N, Karunarathne D and Thilakasiri S (2019), Management of municipal solid waste open dumps immediately after the collapse: An integrated approach from Meethotamulla open dump, Sri Lanka, Waste Management, 95, 227–240.

Jayaweera N, Jayasinghe CL, Weerasinghe SN (2018), Local factors affecting the spatial diffusion of residential photovoltaic adoption in Sri Lanka, Energy Policy, 119, 59–67.

Le Thai-Ha, Chang Y, Taghizadeh-Hesang F. and Yoshino N (2019), Energy insecurity in Asia: A multi-dimensional analysis, Economic Modeling, 83, 84–95.

Mallawarachchi H and Karunasena G (2012), Electronic and Electrical waste management in Sri Lanka: Suggestions for national policy enhancement, Resources, Conservation and Recycling, 68, 2012, 44–53.

Miller D (2011), Selling solar – The diffusion of renewable energy in emerging markets, Earthscan, London, Sterling, VA.

MoMD&E GoSL (2016), Nationally determined contributions (NDCs), Ministry of Mahaweli Development & Environment, Government of Sri Lanka.

Munasinghe M (2019), Sustainable Sri Lanka 2030 Vision and Strategic Path. Presidential Secretariat, Colombo, Sri Lanka.

National Renewable Energies Laboratory (NREL) (2003), Wind Energy Resource Atlas of Sri Lanka and the Maldives,

Premarathne DM, Suzuki N, Ratnayake NP, Kularathne EK (2013), A Petroleum System in the Gulf of Mannar Basin, Offshore Sri Lanka, In: the Proceedings of the Technical Sessions of Geological Society of Sri Lanka, 2013, 9-12 February.

Shah SAA, Zhou P, Walasai GH and Mohein M (2019), Energy security and environmental sustainability index of South Asian countries: A composite index approach, Ecological Indicators, 106, 105507.

Sri Lanka Sustainable Development – A Bill (SLSDB, 2017), Presented by the Minister of National Policies and Economic Affairs, 9th Jan 2017.

Sri Lanka Sustainable Energy Authority (SLSEA) (2018), Sri Lanka Energy Balance 2017, Colombo, Sri Lanka.

Sugathapala T (2020), Multi-criteria assessment of energy sector nationally appropriate mitigation actions in Sri Lanka, IOP Conf. Ser.: Earth & Environmental Sci, 463, 012181.

UN–ESCAP (2019), Policy Perspectives 2019: Sustainable Energy in Asia and the Pacific,. https://www.unescap.org/resources/policy-perspectives-2019-sustainable-energy-asia-and-pacific

United Nations Development Programme (UNDP) (2017), Sri Lanka on Path to 100% Renewable Energy Says a New Joint Report by UNDP and ADB, https://www.undp.org/content/undp/en/home/presscenter/articles/2017/08/16/sri-lanka-on-path-to-100-renewable-energy-says-new-report-by-undp-and-adb.html

Wickramasinghe H (2020). Data on daily electricity demand curves in March, April, and May in Colombo, Sri Lanka (Personal Communication, 12/05/2020)

World Bank and International Finance Corporation (2019), Sri Lanka Energy InfraSAP, Washington, DC 20433: World Bank

10 South Asian Dual Challenges: Energy and Environment

*John Andrew Howe, PhD and
Kankana Dubey*

CONTENTS

10.1 Interregional Relationships, Partnerships, Roles, and Networks 218
10.2 Energy Supply and Demand 220
    10.2.1 Electricity Generation 222
    10.2.2 Energy Consumption by Economic Sector 223
    10.2.3 Energy Mix 224
10.3 Intended Nationally Determined Contributions 226
    10.3.1 Afghanistan 230
        10.3.1.1 Adaptation 230
        10.3.1.2 Mitigation 230
    10.3.2 Bangladesh 231
        10.3.2.1 Adaptation 231
        10.3.2.2 Mitigation 232
    10.3.3 Bhutan 233
        10.3.3.1 Mitigation 234
        10.3.3.2 Adaptation 234
    10.3.4 India 234
        10.3.4.1 Adaptation 235
        10.3.4.2 Mitigation 236
    10.3.5 Maldives 237
        10.3.5.1 Adaptation 238
        10.3.5.2 Mitigation 238
    10.3.6 Nepal 239
    10.3.7 Pakistan 240
        10.3.7.1 Adaptation 241
        10.3.7.2 Mitigation 242
    10.3.8 Sri Lanka 242
        10.3.8.1 Adaptation 243
        10.3.8.2 Mitigation 244
        10.3.8.3 Loss and Damage 244

10.4 Challenges..244
       10.4.1 Energy Security 246
       10.4.2 Energy Equity...247
       10.4.3 Environmental Sustainability of Energy Systems...........................248
10.5 Conclusion 249
References..250

10.1 INTERREGIONAL RELATIONSHIPS, PARTNERSHIPS, ROLES, AND NETWORKS

Taking an optimistic view of South Asia, it can be said that it is an emerging region and global technical hub, characterized as having one of the fastest-growing populations of educated young people, emerging trends in higher living standards, increased industrial energy demand, multiple coexisting religious faiths, and increasing purchasing power.

South Asia's economic momentum since 2016 is unprecedented, with the entire region growing faster than China. The Global Competitiveness Report 2017–18, published by the World Economic Forum reports that overall competitiveness has improved across most countries in this region, with significant changes in Bhutan, Nepal, Pakistan, and Bangladesh (Schwab, 2017).

Today the world is far more interconnected than ever before; truly globalized, the spread of COVID-19 is an unpleasant testimony of the globalized world. The slogan "Think Local, Act Global" has changed the dynamics of global trade, politics, regional cooperation, consumer consumption patterns, etc. However, within this globalized world, there still exist many regional entities that operate through formidable cooperation, using their strength to interact with the rest of the world. One such successful regional cooperation is the European Union, comprised of all the countries on the European continent. Likewise, there are several other regional organizations, such as the African Union (AU), Association of Southeast Asian Nations (ASEAN), Council of Europe (COE), League of Arab States (LAS), Organization of American States (OAS), Pacific Islands Forum (PIF), and South Asian Association for Regional Cooperation (SAARC). These organizations can also be thought of as continental unions, designed to facilitate geographical integration between countries. They create trade blocs, boost regional economies, and encourage peace and harmony by developing legal frameworks to protect the interests of the member states. The EU is, by far, the single most successful regional bloc, possibly due to political will and commitment. In comparison to the EU, the regional integration of developing countries is weak, mainly due to the lack of cooperation and commitment, and the differences in economic and natural resources. In this chapter, we review the South Asian Association for Regional Cooperation (SAARC) countries' regional perspectives on tackling regional energy and environmental challenges, interdependencies, and strategies for a cohesive future of sustainable development.

SAARC (Figure 10.1) was established in 1985, by seven South Asian nations; Afghanistan joined the cooperation in 2005. The map in Figure 10.1 shows population estimates for 2020 (United Nations Department of Economic and Social Affairs,

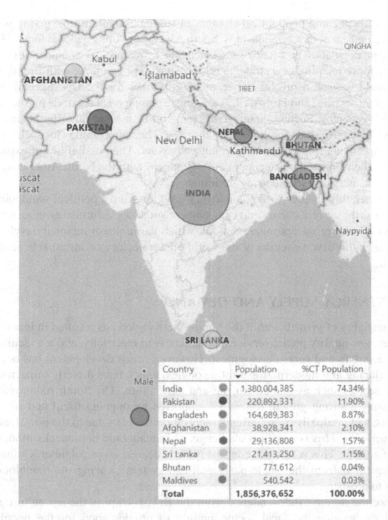

Country		Population ▼	%CT Population
India	●	1,380,004,385	74.34%
Pakistan	●	220,892,331	11.90%
Bangladesh	●	164,689,383	8.87%
Afghanistan	●	38,928,341	2.10%
Nepal	●	29,136,808	1.57%
Sri Lanka	●	21,413,250	1.15%
Bhutan	●	771,612	0.04%
Maldives	●	540,542	0.03%
Total		**1,856,376,652**	**100.00%**

FIGURE 10.1 SAARC member nations and 2020 population estimates. (Source: United Nations Department of Economic and Social Affairs, Population Division, 2019.)

Population Division, 2019), indicated by the sizes of the bubbles; the combined population in the region was more than 1.85 billion people, led by India. The economic areas initially designated for cooperation were only five sectors: agriculture, rural development, telecommunications, meteorology, and health and population. Additional areas of transport, postal services, and scientific and technological cooperation were added later, along with sports, arts, and culture. Paradoxically, collaboration within the South Asian countries continues to strive to exist as significant divisions remain throughout the region, with the history of conflicts preventing security and trust. These are the single most reasons that have slowed the economic progress of the region and limited its potential. Unlike the EU, the development of SAARC has been slow and has not fulfilled its objective to foster and strengthen cooperation between the countries of South Asia.

With the ongoing geopolitical changes in strategic relationships, giving way for the development of new relationships, South Asia is becoming a global focus. Having a growing population and scope for tremendous economic development, it is on the frontier for developing new strategic relationships. For example, with Brexit, South Asia could become a pivotal choice of trade partner for the UK. Specializing in agriculture, textiles, and IT, SAARC can bring a complementary trade partnership to the UK's economy. Further, South Asia's burgeoning market can offer business and economic opportunities for the UK's industries. The opportunities for such a strategic tie-up are not limited to the UK, but go beyond. The possibilities of expanding trade relationships to boost economic development with the North American, Latin American, and African countries are immense.

However, this chapter does not discuss the region from a political standpoint, but will assess the current energy and environment situation, culminating in recommendations to achieve an optimistic outlook which can unleash regional development and launch this diverse region of the world on a trajectory of sustainable economic growth.

10.2   ENERGY SUPPLY AND DEMAND

The inequality of growth within the South Asian region has resulted in income disparities, low-quality public services, poor access to electricity, and a widening gap between urban and rural development. However, much development has occurred in developing and modernizing public policies, which have directly impacted economic sectors such as manufacturing and agriculture. The South Asian region is witnessing a strong growth trajectory supported by ongoing fiscal and structural reforms, and gradually overcoming structural bottlenecks due to the prevalent rigid bureaucracies. This is a critical challenge, as bureaucratic bottlenecks hinder economic growth. This is in stark contrast to their nearest large neighbor, China, who has benefitted from the single-party political system clearing the roadblocks for China to prosper.

Since the region's independence from the British empire, the agriculture sector singularly became the focal sector, mainly to provide food for the population. Historically, the agriculture sector was heavily subsidized with energy, water, and land allocation. At this point in the region's economic maturity, in order to tap the potential growth, investment should be redirected toward the manufacturing and industrial sectors. These sectors have consistently suffered from a lack of infrastructure, power outages, and opposing fiscal policies. It is thus imperative that SAARC countries evaluate their current economic policy to support the productive growth of the manufacturing sector by incorporating sustainable practices that could support economic progress and reducing dependencies on fossil fuel. To develop the other economic sectors, the regional governments first need to create an uninterrupted energy supply to meet energy demand, develop financial tools, and build infrastructure to encourage local and international private investment. Perhaps most importantly, regressive electricity subsidies prevalent across the region need to be reduced or eliminated. These measures could lead to higher productivity growth and promote energy efficiency across various sectors of the economy. For the region to propel

TABLE 10.1

Potential Energy Generation Capacity and Reserves in South Asia

	Generation Capacity (MW)			Reserves			
Country	Wind	Utility-Scale Solar	Hydropower	Coal (million tons)	Oil (million barrels)	Natural Gas (trillion ft³)	Biomass (million tons)
Afghanistan	67,000	220,000	23,000	440		15	23
Bangladesh			4,000	882	12		
Bhutan	4,825		263,000	4		8	27
India	102,778	750,000	150,000	90,085	57,000	39	139
Maldives							
Nepal	3,000		730,000				27
Pakistan	131,800	169,000	60,000	17,550	324	33	
Sri Lanka	24,000		21,000				12
Total	333,403	1,139,000	1,251,000	108,961	57,336	95	228

Source: Alam et al., 2019.

itself on the path of economic and sustainable progress, an uninterrupted supply of energy is pivotal.

According to Alam et al. (2019), the eight SAARC member states are deficient in liquid fuel reserves, with the exception of India which has significant oil reserves. Table 10.1 compares the renewable energy generation capacity and available fossil fuel reserves in place. The entire region is highly dependent on crude oil and/or refined petroleum product imports. Oil import risks, driven largely by the fluctuation of global supply and prices, weigh heavily upon SAARC member state economies and development activities. Hence, nearly one quarter of the global population faces substantial energy insecurity.

This section provides detailed views of the energy mix of SAARC countries. First, we will consider the fuel mix for electricity generation. This is an important view, as clean and reliable electricity is of paramount importance for economic development. Additionally, electricity generation is a large contributor to GHG emissions. Next, we decompose energy consumption by energy source and economic sector. Finally, we analyze the trends and composition of total primary energy supply.

Due to a lack of reliable and consistent data, Afghanistan, Bhutan, and the Maldives will be excluded from the analysis. In Bhutan, nearly 100% of the country's electricity is generated from run-of-the-river hydropower (National Environment Commission, Royal Government of Bhutan, 2015). According to the Maldives' Intended Nationally Determined Contributions (Ministry of Environment and Energy, Government of Maldives, 2015), the 2011 energy mix was approximately 4% LPG, 0.1% solar photovoltaic (PV), and the remainder from imported petroleum products (diesel, petrol, jet A1, and kerosene).

10.2.1 ELECTRICITY GENERATION

Table 10.2 shows the total electricity consumption for each country in South Asia in 2019, except Maldives. India, with the largest population, consumes more electricity than six times all the other countries put together.

The left pane of Figure 10.2 shows the trend in electricity supply and the share of fossil fuel and renewable sources. As electricity supply has increased by nearly 20%, the share of renewables and nuclear power in the region has actually decreased from 18.4% to 17.6%. Overall, a full 80% of electricity in these five SAARC member states is generated from fossil fuels. On the right pane, we see boxplots for the shares of fossil fuel and renewables, demonstrating the high amount of variability in the region. Nepal does not burn any fossil fuels, but relies on hydropower (64%) and electricity imports. Bangladesh, on the other hand, generates 98% of electricity by burning fossil fuels, with 80% from natural gas.

TABLE 10.2
2019 Electricity Consumption by Country

Country	Electricity Consumption (billion kWh)
India	1,137
Pakistan	92
Bangladesh	54
Sri Lanka	13
Afghanistan	6
Nepal	5
Bhutan	2

Source: Central Intelligence Agency (n.d.).

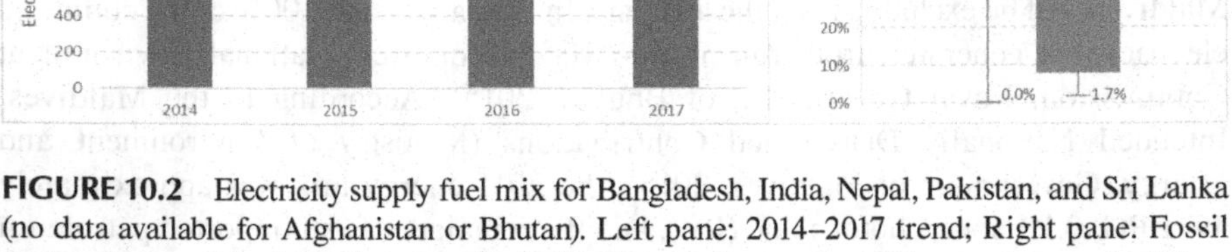

FIGURE 10.2 Electricity supply fuel mix for Bangladesh, India, Nepal, Pakistan, and Sri Lanka (no data available for Afghanistan or Bhutan). Left pane: 2014–2017 trend; Right pane: Fossil fuel and renewable shares boxplots for 2017. (Source: International Energy Agency (n.d.).)

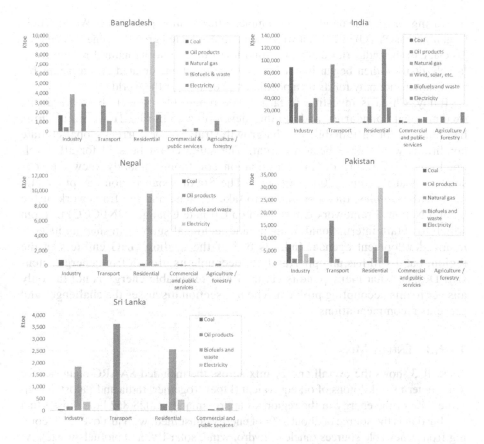

FIGURE 10.3 2017 energy consumption by fuel type by economic sector. (Source: International Energy Agency (n.d.).)

10.2.2 ENERGY CONSUMPTION BY ECONOMIC SECTOR

The graphs in Figure 10.3 present the stark reality that most of the energy consumed in the residential sector comes from biofuels and waste. This does not truly solve the issue of energy access but instead leads to the concern that energy access comes at the cost of the environment. Similarly, most SAARC member nations rely on coal combustion to provide energy for industry. Given the depleting coal reserves across the region, neither is this sustainable, nor is coal a clean and efficient fuel. Burning coal contributes heavily to greenhouse gas (GHG) emissions, which are a major concern for the international community and a big threat to the South Asian region. The *Climate Risk Index* published by Germanwatch is a testament to the importance of mitigating climate change. In the most recent report (Eckstein et al., 2019), four of the eight SAARC member nations were in the top five most vulnerable countries globally.

Unleashing economic activity without accounting for long-term sustainability increases the energy intensity of the economy, hindering energy security and increasing energy demand. For example, after China joined the World Trade Organization (WTO) in 2000, it witnessed unprecedented growth in energy demand. To cater to the industrial demand, China first dipped into its natural resources — largely coal — then began importing oil. These actions created an unsustainable environment not only for its own population, but also for the world.

Today, China is investing in large-scale renewable projects both within its own national boundaries, and in other developing countries. However, if China had planned for sustainability in its growth strategy, the current global climate conditions would have been different. The same could be said for all developed countries during the industrialization era, though nobody knew the consequences at the time (200 years ago). The South Asian region has pragmatic historical examples, the opportunity to take lessons, and the framework of the United Nations Framework Convention on Climate Change (UNFCCC) through which to obtain international financial and technical support in shaping its economic development agenda. India is one of the leading GHG emitters in the region, but it still has an opportunity to decouple its growth from energy intensity. Despite what many pundits claim, 100% renewable energy is not the only answer to this decoupling problem. The last section discusses the challenges and presents recommendations.

10.2.3 ENERGY MIX

Table 10.3 shows the overall energy mix across the included SAARC countries for 2017, in terms of kilotons of oil equivalent (ktoe). Together, India and Pakistan consume 94% of the energy in the region — approximately 1,028,852 ktoe. Fossil fuel combustion is the source of about 74% of energy consumed, with just over 25% coming from renewable sources (nuclear, hydro, wind, solar PV, and biofuels/waste). As with electricity supply, there is a substantial amount of variability in the energy mix. One commonality is that most renewable energy generated by these five countries is from biofuels/waste. While this is clearly an inexpensive and readily available fuel, combustion of these sources pollutes egregiously. This is a reminder of the harsh reality that not all renewable sources are clean. Three-quarters of Nepal's energy is generated renewably, while this share plummets to 23.5% for India. The most important fossil fuels in the region are coal (37%), and crude oil (26%); combined, these two fuels account for 689,126 ktoe — about 62% of all energy. Bangladesh and Pakistan are the most reliant on natural gas, which represents about 58% and 25% of their energy mixes, respectively.

The share of renewables in the total primary energy supply over the four years from 2014 to 2017 has followed a similar trend as seen for electricity generation. Figure 10.4 shows that as total energy consumed has increased in the region by about 7%, the share of renewables has fallen from 26.4% to 25.7%, with increased fossil fuel consumption making up the difference. Where the actual energy generated from renewables has increased, most has been increased combustion of biofuels/waste. The sole contrary example is India, which increased clean renewable energy (wind, solar, etc.) from 4.2 million toe (mtoe) to 7.5 mtoe — an increase of 75%.

TABLE 10.3
2017 Energy Supply Mix, in Kilotons of Oil Equivalent

Source	Bangladesh	India	Nepal	Pakistan	Sri Lanka	Total
Coal	984	390,944	791	10,750	1,510	**404,979**
	(2.5%)	(42.3%)	(5.9%)	(10.3%)	(12.6%)	**(37.0%)**
Crude oil	1,591	265,762	0	15,309	1,485	**284,147**
	(4.0%)	(28.7%)	(0%)	(14.7%)	(12.3%)	**(26.0%)**
Oil products	4,237	0	2,298	12,528	4,033	**23,096**
	(10.7%)	(0.0%)	(17.0%)	(12.0%)	(33.5%)	**(2.1%)**
Natural gas	23,071	51,021	0	26,036	0	**100,128**
	(58.4%)	(5.5%)	(0%)	(24.9%)	(0%)	**(9.2%)**
Fossil Fuels	**29,883**	**707,727**	**3,089**	**64,623**	**7,028**	**812,350**
	(75.6%)	**(76.6%)**	**(22.9%)**	**(61.9%)**	**(58.4%)**	**(74.3%)**
Nuclear	0	9,991	0	2,574	0	**12,565**
	(0%)	(1.1%)	(0%)	(2.5%)	(0%)	**(1.1%)**
Hydro	90	12,193	398	2,401	346	**15,428**
	0.2%)	(1.3%)	(3.0%)	(2.3%)	(2.9%)	**(1.4%)**
Wind, solar, etc.	16	7,479	1	247	44	**7,787**
	(0%)	(0.8%)	(0%)	(0.2%)	(0.4%)	**(0.7%)**
Biofuels/waste	9,534	187,138	9,778	34,568	4,608	**245,626**
	(24.1%)	(20.2%)	(72.5%)	(33.1%)	(38.3%)	**(22.5%)**
Renewables	**9,640**	**216,801**	**10,177**	**39,790**	**4,998**	**281,406**
	(24.4%)	**(23.5%)**	**(75.5%)**	**(38.1%)**	**(41.6%)**	**(25.7%)**
Electricity	0	-137	222	48	0	**133**
	(0%)	(0%)	(1.6%)	(0%)	(0%)	**(%)**
Total Primary	**39,523**	**924,391**	**13,488**	**104,461**	**12,026**	**1,093,889**
Energy Supply	**(3.6%)**	**(84.5%)**	**(1.2%)**	**(9.5%)**	**(1.1%)**	

Source: International Energy Agency (n.d.).

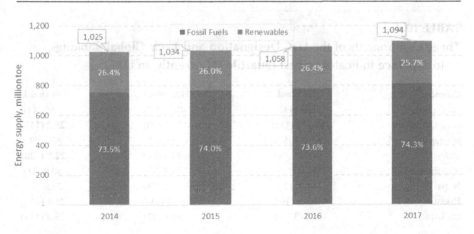

FIGURE 10.4   Energy supply mix in million toe from 2014 to 2017. (Source: International Energy Agency (n.d.).)

10.3 INTENDED NATIONALLY DETERMINED CONTRIBUTIONS

With the exception of Afghanistan, the countries of SAARC ratified their membership in the UNFCCC between 1992 and 1995. The Maldives was the first to ratify, followed by India and Sri Lanka. Afghanistan was the last to ratify its membership in 2002, though they signed the treaty more than a decade earlier. In this section, we will evaluate and compare each country's Intended Nationally Determined Contribution (INDC) along several dimensions. Before discussing the INDCs, however, it is instructive to provide contextual information about the region. The most relevant considerations are economic development, human development, and climate change vulnerability. While there are differences among the countries, there are several relevant commonalities.

Four of the SAARC member states are currently designated as Least Developed Countries (LDCs): Afghanistan, Bangladesh, Bhutan, and Nepal (United Nations Department of Economic and Social Affairs, 2018). A country is classified by the United Nations as an LDC if it meets specific criteria relating to income (Gross National Income per capita; GNI), human assets (Human Assets Index; HAI), and economic vulnerability (Economic Vulnerability Index; EVI). These three indicators are chosen for the manner in which they reflect long-term structural handicaps. Table 10.4 shows the three indices from 2018 for the South Asian countries. For GNI per capita (2014–2016 average), Afghanistan, Bangladesh, and Nepal are in the lowest quartile globally. Considering the UN's HAI, only Afghanistan and Pakistan fall in the lower quartile. Finally, Bangladesh, India, Pakistan, and Sri Lanka fall in the lower quartile of EVI.

The United Nations Development Programme gives a slightly different view with their Human Development Index (HDI), which is computed from life expectancy, education, and national income indicators. Table 10.5 lists the HDI and Gross National Income per capita from the Human Development Report Office (2019). According to this report, all the SAARC countries but Sri Lanka are ranked in the lower half of the world in terms of HDI. In fact, Afghanistan, Pakistan, Nepal,

---

TABLE 10.4

Three Components of the LDC Designation and Their Global Rankings; Bold Typeface Indicates Lowest Quartile (*Currently an LDC)

Country	GNI (Rank)	HAI (Rank)	EVI (Rank)
Afghanistan*	**$633 (132)**	**48.4 (126)**	39.3 (41)
Bangladesh*	**$1,274 (111)**	73.2 (90)	**25.2 (111)**
Bhutan*	$2,401 (88)	72.9 (91)	36.3 (56)
India	$1,591 (99)	74.2 (89)	**22.9 (126)**
Maldives	$9,200 (33)	91.4 (49)	50.9 (21)
Nepal*	**$745 (125)**	71.2 (95)	28.4 (95)
Pakistan	$1,502 (102)	**56.7 (114)**	**21.9 (128)**
Sri Lanka	$3,773 (73)	91.9 (47)	**25.0 (114)**

Source: United Nations Department of Economic and Social Affairs (2018).

TABLE 10.5

Human Development Index (HDI), Gross National Income (GNI), and Their Global Rankings

Country	HDI	Rank	GNI	Rank
Afghanistan	0.496	170	$1,746	171
Bangladesh	0.614	135	$4,057	141
Bhutan	0.617	134	$8,609	111
India	0.647	129	$6,829	124
Maldives	0.719	104	$12,549	87
Nepal	0.579	147	$2,748	160
Pakistan	0.56	152	$5,190	135
Sri Lanka	0.78	71	$11,611	95

Source: Human Development Report Office (2019).

Bangladesh, and Bhutan are all in the lower third decile. Along with economic and human development, another common factor is high vulnerability to the impacts of climate change.

Table 10.6 shows Climate Risk Index (CRI) scores for all countries, published by Germanwatch, covering 2015 through 2018 (Eckstein et al. (2017–2019) and Kreft et al. (2016)). As of the most recent report, Afghanistan, India, Nepal, and Sri Lanka were all in the most vulnerable second decile. As can be seen in the left pane of Figure 10.5, from 2015 (2016 for the Maldives, when the data start) to 2018, three South Asian countries' CRI rankings increased — Afghanistan, Maldives, and Sri Lanka all demonstrated higher vulnerability. Sri Lanka's CRI ranking skyrocketed from 98th

TABLE 10.6

CRI Scores (and Rankings) from Germanwatch Global Climate Risk Index Reports for 2015–2018 (published with a two year lag); a Larger CRI Indicates Lower Vulnerability

Country	2015	2016	2017	2018
Afghanistan	43.3 (28)	39.3 (24)	37.3 (26)	36.0 (24)
Bangladesh	46.2 (35)	27.0 (13)	16.0 (9)	85.5 (98)
Bhutan	76.2 (87)	67.2 (75)	116.0 (124)	125.0 (135)
India	15.3 (4)	18.3 (6)	22.7 (14)	18.2 (5)
Maldives	-	109.5 (120)	116.0 (124)	103.3 (118)
Nepal	51.3 (42)	29.5 (14)	10.5 (4)	29.7 (20)
Pakistan	28.2 (11)	50.8 (40)	43.2 (33)	87.8 (100)
Sri Lanka	86.0 (98)	11.5 (4)	9.0 (2)	19.0 (6)

Source: Eckstein et al. (2017–2019) and Kreft et al. (2016).

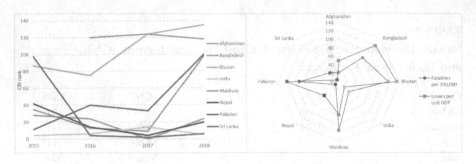

FIGURE 10.5   Climate vulnerability rankings for 2018. Left pane: overall CRI; Right pane: fatalities per 100,000, and losses per unit GDP. (Data source: Eckstein et al. (2019).)

(near the global median) in 2015 to 4th in a single year! The remaining countries show decreased relative vulnerability over the same timeframe, with Pakistan's vulnerability decreasing nearly as much as Sri Lanka's increased (global rank of 11th down to 100th).

The right pane of Figure 10.5 shows, as per the 2020 Global Climate Risk Index report, the rankings for the two primary determinants of CRI for each country: *Fatalities per 100,000 people* (filled circles), and *Losses per unit GDP%* (filled squares). This is a useful view of vulnerability, as it allows us to see the relative importance of each factor in determining a country's vulnerability index. India, Maldives, and Sri Lanka all exhibit a larger financial impact of climate vulnerability — the loss rankings for these countries are higher than the fatality rankings. This is perhaps because these three countries have the highest proportion of coastal areas. The remaining countries have higher vulnerabilities due to fatalities. Afghanistan, Nepal, and Sri Lanka are all ranked very high for fatalities: 21st, 10th, and 29th, respectively.

Table 10.7 lists some high-level characteristics of the Intended Nationally Determined Contributions (INDCs) of the eight SAARC member states. In the remainder of this section, we will evaluate and compare them in more detail.

TABLE 10.7
INDC Characteristics by Country; the First Column Indicates the Year of UNFCCC Membership Ratification (*Currently an LDC)

Country	Base Year	Mitigation Type	GHG Target	Non-GHG Target	GHG Covered
Afghanistan (2002)*	2005	GHG target	13.6% reduction in GHG emissions by 2030 compared to a Business as usual (BAU) scenario, conditional on external support	Baseline scenario target	CO_2, CH_4, N_2O
Bangladesh (1994)*	2011	GHG target	5% (unconditional) to 15% (conditional) reduction in GHG emissions by 2030 compared to the BAU levels	Renewable energy target	CO_2, CH_4, HFCs, PFCs, SF_6
Bhutan (1995)*	-	GHG & non-GHG targets	Remain carbon neutral where GHG emission will not exceed sequestration by forests, estimated at 6.3 million tons of CO_2	Fixed-level target	CO_2, CH_4, N_2O
India (1993)	2005	GHG & non-GHG targets	33% to 35% reduction in the emissions intensity of its GDP by 2030 compared to 2005 level	Intensity target	Not specified
Maldives (1992)	2011	GHG target	10% (unconditional) up to 24% (conditional) reduction in GHG emissions for the year 2030 compared to the BAU scenario	Baseline scenario target	CO_2, CH_4
Nepal (1994)*	2015	Non-GHG target and actions	N/A	N/A	Not specified
Pakistan (1994)	2015	GHG target	Up to 20% reduction in GHG emissions by 2030 compared to the projected GHG emissions, conditional	Baseline scenario target	Not specified
Sri Lanka (1993)	2010	GHG target	4% (unconditional) to 20% (conditional) reduction in GHG emissions from the energy sector, and 3% (unconditional) to 10% (conditional) from other sectors (transport, industry, forests and waste) by 2030 compared to the BAU scenario	Baseline scenario target	CO_2, CH_4, N_2O

Sources: Country-specific INDCs, and United Nations Department of Economic and Social Affairs (2018).

10.3.1 AFGHANISTAN

HDI (Rank)	GNI/ Capita	CRI (Rank)	EVI (Rank)	HAI (Rank)	Conditional
0.496 (170)	$1,746	36.0 (24)	39.3 (41)	48.4 (126)	True

GHG Target: 13.6% reduction in GHG emissions by 2030 compared to a BAU scenario, conditional on external support. (Source: Islamic Republic of Afghanistan (2015).)

As an LDC with the lowest GNI per capita in the region, HDI in the lowest decile, CRI in the top 15th percentile, and a heavily agrarian society, Afghanistan is uniquely vulnerable to climate change. According to Afghanistan's INDC, the country is already beginning to experience the initial adverse impacts of climate change. An increase of 0.6°C in the country's mean annual temperature since 1960 has been documented, and springtime rainfall has decreased by approximately 40.5 mm. Climate change projections suggest that Afghanistan's environment will experience more substantial changes over the rest of the century. Anticipated detrimental impacts include increased incidence of food- and water-borne diseases, enhanced social inequality, and decreased food security.

10.3.1.1 Adaptation

The country is grappling with numerous specific challenges in addressing climate change through adaptation. These challenges include funding gaps, lack of expertise, lack of reliable long-term historical climate data, weak public awareness about environmental issues, and security. As part of their National Adaptation Plan, Afghanistan has identified several high-level key actions, which will require external funding of $10.785 billion over ten years. The actions requiring the most funding and support are:

- restoration and development of irrigation systems to increase irrigated agricultural land
- planning for proper watershed management and promotion through community-based natural resources management
- regeneration of at least 40% of existing degraded forests (232,050 ha) and rangeland areas (5.35 Mha)

10.3.1.2 Mitigation

Afghanistan is one of the lowest GHG emitters, with per capita emissions at 0.3 metric tons in 2010. Emissions in 2005 were 28,758 metric tons CO_2 equivalent (CO_2e). This is projected to increase to over 61,000 metric tons by 2030, under the Business as Usual (BAU) scenario. Figure 10.6 displays the country's emissions by sector and gas, for 2005 and forecast to 2030. The majority of emissions from 2005 were CO_2 from land-use change and forestry (LUCF). The balance of emissions is projected to shift from 13% for energy generation to 20%, representing mostly a decrease in emissions from LUCF.

To mitigate this 212% increase in emissions, Afghanistan's INDC includes several proposed mitigation contributions, which are conditional upon financial and technical support being met, valued at $6.62 million per year from 2020. These contributions include building and transport energy efficiency efforts, clean and

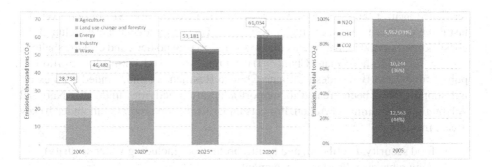

FIGURE 10.6 GHG emissions by sector and gas. Left pane: Projected increase to 2030. Right pane: Composition by GHG. (Source: Islamic Republic of Afghanistan (2015).)

renewable electricity generation technology, forest-based carbon sequestration, and low-carbon agriculture.

10.3.2 BANGLADESH

HDI (Rank)	GNI/ Capita	CRI (Rank)	EVI (Rank)	HAI (Rank)	Conditional
0.614 (135)	$4,057	85.5 (98)	25.2 (111)	73.2 (90)	True

GHG Target: 5% (unconditional) to 15% (conditional) reduction in GHG emissions by 2030 compared to the BAU levels. (Source: Ministry of Environment and Forests, Government of the People's Republic of Bangladesh (2015).)

Bangladesh is currently classified by the United Nations as an LDC, as it was when its INDC was submitted to the UNFCCC. In 2017, Bangladesh was in the top 10 most climate vulnerable countries (ranked 9), but in the most recent report, the country's ranking has dropped into the lower half of all countries. Among the South Asian countries, it has the 3rd lowest GNI per capita. According to Bangladesh's Ministry of Environment and Forests, emissions are less than 0.35% of global emissions. Bangladesh is situated in the flood plain of several major rivers, and sits at the apex of the Bay of Bengal; consequently, the country is particularly susceptible to extreme weather events, which are increasing in frequency and severity due to global warming (Dow and Downing, 2011). Failure to act ambitiously and decisively to combat climate change could cost Bangladesh an annual loss of 2% of GDP by 2050 and 9.4% of GDP by 2100. The country's INDC details a two-pronged strategy for handling climate change: primarily adaptation, followed by mitigation.

10.3.2.1 Adaptation

Weather events driven by climate change, such as extreme temperatures, erratic rainfall, rising sea levels, and ocean acidification, have already had serious negative consequences for the lives of millions of Bangladeshis. These events are gradually canceling out 30 years of substantial socioeconomic development and jeopardizing future improvements. Hence, the focus of the government's activities is on bolstering resilience against these impacts. The country has invested over $10 billion over the last three decades to enhance climate resilience. These investments have included the construction of flood management embankments, coastal polders, and cyclone shelters.

The Bangladesh Climate Change Trust Fund, endowed with $400 million of public funds, and Bangladesh Climate Change Resilience Fund, funded with external support, have been created to support adaptation activities under the National Adaptation Programme of Action. Projects under the action program fall under 10 key areas, including:

1. food security, livelihood, and health protection (including water security)
2. comprehensive disaster management
3. coastal zone management (including salinity intrusion control)
4. flood control and erosion protection
5. building climate-resilient infrastructure

As of June 2015, more than 230 projects had been funded, with over 40 implemented. An estimated US$42 billion between 2015 and 2030 will be needed to fully fund all identified adaptation measures.

10.3.2.2 Mitigation

Bangladesh's contribution to climate change mitigation covers the power, transport, and industry sectors, which are forecast (under the BAU scenario) to represent 69% of the country's emissions by 2030 — -234 $MtCO_2e$. Key mitigation objectives, which are conditional upon international finance, technology transfer, and capacity-building support include those shown in Table 10.8. The key mitigation measures in these three sectors are projected to cost almost $27 billion. Possible mitigation actions in other sectors (also conditional), the costs of which have not yet been quantified include incentivizing the adoption of more efficient household stoves, incentivizing rainwater harvesting, increasing agricultural mechanization, and reforestation.

TABLE 10.8
Key Mitigation Measures

Sector	Objectives
Energy	100% of new coal-based power plants use super-critical technology by 2030
	400 MW of wind-generating capacity by 2030
	1000 MW of utility-scale solar power plant
Transport	To achieve a shift in passenger traffic from road to rail of up to around 20% by 2030 compared to the business as usual scenario
	15% improvement in the efficiency of vehicles due to more efficient running
Industry	10% energy consumption reduction in the industry sector compared to the business as usual scenario

Source: Ministry of Environment and Forests, Government of the People's Republic of Bangladesh (2015).

10.3.3 BHUTAN

HDI (Rank)	GNI/ Capita	CRI (Rank)	EVI (Rank)	HAI (Rank)	Conditional
0.617 (134)	$8,609	125.0 (135)	36.3 (56)	72.9 (91)	True

GHG Target: Remain carbon neutral where GHG emission will not exceed sequestration by forests, which is estimated at 6.3 million tons of CO₂. (National Environment Commission, Royal Government of Bhutan (2015).)

Bhutan, a small developing country in a mountainous region, has been committed to remaining carbon neutral since 2009. It is estimated that the forests of Bhutan, which cover approximately 33,000 km² (70.5% of the country's land area) have the capacity to sequester 6.3 million tons of CO_2. The country is an LDC with a Human Development Index in the lowest third globally, though with the third highest national income per capita among SAARC member states. Bhutan's status as an LDC can be attributed more to economic vulnerability (ranked 51) than to the Human Assets Index (ranked 91). The country has the lowest Climate Risk Index out of all the SAARC countries. Nearly 100% of the country's electricity is generated from run-of-the-river hydropower. The electrification rate in urban areas is almost 100%, while almost 95% of rural residents have electricity. Bhutan exports hydropower to neighboring countries, which is a substantial source of revenue to the country and offsets more than 4 million tons of CO_2 equivalent emissions.

Bhutan is lightly industrialized, and agriculture is one of its largest sources of emissions. However, as industrialization spreads, emissions are increasing rapidly. Between 2000 and 2013, emissions in the energy, industry, and waste management industries all more than doubled. The waterfall charts in Figure 10.7 show the change in emissions in these sectors, and how that has reduced the net carbon sequestration available by Bhutan's forests from 4.7 million tons of CO_2e to 4.1 by 2013.

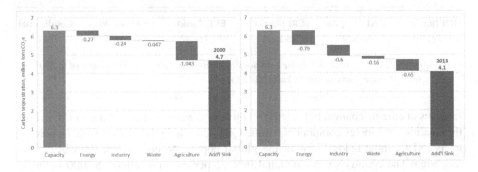

FIGURE 10.7 Waterfall charts for 2000 (left) and 2013 (right) showing GHG emissions and sequestration. (Source: National Environment Commission, Royal Government of Bhutan (2015).)

As an LDC, Bhutan has an imperative to develop economically and is pursuing a sustainable development path; the country already has several national policies in place to support climate change mitigation and adaptation efforts. The Economic Development policies of 2010 and 2015 (draft) include several measures promoting green growth in industry. In addition, the INDC details a host of potential adaptation and mitigation strategies, which depend on receiving sufficient external financial and technical support. Some of the highest priorities are listed here.

10.3.3.1 Mitigation

The following are mitigation measures for Bhutan:

1. sustainable forest management and conservation of biodiversity to ensure sustained environmental services
2. promotion of low-carbon transport system
3. minimize GHG emission through the application of zero-waste concepts and sustainable waste management practices
4. promote a green and self-reliant economy toward carbon-neutral and sustainable development
5. promote clean renewable energy generation

10.3.3.2 Adaptation

The following are adaptation measures for Bhutan:

1. increase resilience to the impacts of climate change on water security through Integrated Water Resource Management approaches
2. promote climate-resilient agriculture to contribute toward food and nutrition security
3. sustainable forest management and conservation of biodiversity to ensure sustained environmental services
4. strengthen resilience to climate change-induced hazards
5. minimize climate-related health risks

10.3.4 INDIA

HDI (Rank)	GNI / Capita	CRI (Rank)	EVI (Rank)	HAI (Rank)	Conditional
0.647 (129)	$6,829	18.2 (5)	22.9 (126)	74.2 (89)	True

GHG Target: 33% to 35% reduction in the emissions intensity of its GDP by 2030 compared to 2005 level. (Source: Government of India (2015).)

In terms of climate change, India is the 5th most vulnerable country in the world; in both the fatalities and losses components of the CRI, India is in the highest fifth. According to its INDC, India holds 17.5% of the global population on 2.4% of the world's surface area. The country's gross national income per capita of almost $7,000 is the 4th highest in the region, but 124th in the world; 30% of the world's poor are in India. As with Bhutan, environmental protection is built into the foundation of the government; the constitution (Article 48-A) states that, *"The State shall endeavour to protect and*

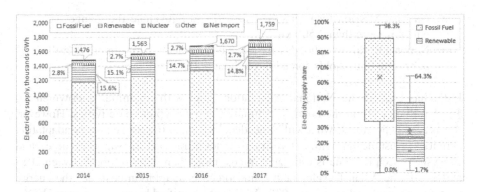

FIGURE 10.8 Forecast growth in population and electricity demand. (Source: Government of India (2015).)

improve the environment and to safeguard the forests and wildlife of the country". India has a voluntary goal to reduce the country's emissions intensity (with respect to GDP) by 20–25% by 2020, relative to 2005. By 2010, a 12% reduction had been realized; for which the United Nations Environment Programme recognized the country as being on course to achieving its goal. Overall, preliminary estimates suggest that at least $2.5 trillion will be required to meet India's climate change actions between now and 2030.

The population of India is expected to grow by 25% by 2030, from 1.2 billion (2014) to 1.5 billion. As shown in Figure 10.8, this growth is forecast to include a substantial shift toward more urbanization. Over almost the same time horizon, electricity demand is expected to more than double from about 650 GWh per capita (2012) to almost 1,700.

10.3.4.1 Adaptation

An Asian Development Bank Study projected that annual climate change-induced economic losses in India could climb to 1.8% of GDP by 2050. Indeed, India is ranked 19 (in the highest decile) for the CRI component of Losses per unit GDP. The Indian subcontinent is among the world's most disaster-prone areas, with nearly 85% of India being vulnerable to at least one hazard. The strengthening impact of climate change on natural disasters has been further amplified due to the predominance of poverty. Twenty-three Indian states/territories — 45.5 million hectares of land — are subject to flooding; during 1996 and 2005, the average annual cost of flood damage was more than $750 million.

India's expenditures for critical adaptation programs was approximately 2.8% of GDP in 2009–2010. Out of the eight national climate change missions, several are focused on adaptation in the agriculture, water, Himalayan ecosystem, and forestry sectors. Some sample programs/strategies include:

- National Mission on Sustainable Agriculture: aims at enhancing food security and protecting resources such as land, water, biodiversity, and genetics
- National Mission for Sustaining the Himalayan Ecosystem: addresses important issues concerning Himalayan Glaciers and the associated hydrological and ecological systems

- National Mission for Clean Ganga: seeks to rejuvenate the Ganges river, which is more than 2,500 km long
- National Rural Livelihoods Mission: has the objective to support 70 million poor rural households spread throughout 600,000 villages

India has created the National Adaptation Fund with an initial endowment of $55.6 million to fund the country's climate change adaptation activities. However, this is a small portion of what will be needed. Implementing the necessary adaptation activities in just the agriculture, forestry, fisheries infrastructure, water resources, and ecosystems sectors alone is estimated to cost more than $200 billion between 2015 and 2030. The Asian Development Bank forecasts adaptation costs for the energy sector to be nearly $8 billion in 2030. While not quantified by project in their INDC, the government of India is anticipating that this will be heavily financed from external sources.

10.3.4.2 Mitigation

India has several programs in place aimed at mitigating global climate change in the energy sector. These include scaling up renewable energy generation (175 GW target), upgrading coal plants to supercritical combustion, and increasing energy efficiency. For example, the National Mission for Enhanced Energy Efficiency (NMEEE) has the goal to strengthen the market for energy efficiency; the program expects to avoid 19,598 MW of capacity additions and achieve fuel savings of around 23 million tons per year at its full implementation stage. Between 2005 and 2012 NMEEE programs have already achieved 10,000 MW avoided generation capacity. Lighting has been a focus, with the share of compact fluorescent lamps rising from 8% in 2005 to 37% in 2005. India has also launched a program to replace all incandescent lamps with LEDs; this is expected to save up to 100 billion kW annually. The country's INDC documents eight major strategies for focused climate change mitigation:

- clean and efficient energy system
- enhancing energy efficiency in industries
- developing climate-resilient urban centers
- promoting waste to wealth conversion
- safe, smart, and sustainable green transportation network
- planned afforestation
- abatement of pollution
- citizen and private sector contribution to combating climate change

The source of the majority of the funds for the programs implemented under these strategies is budgetary. The government is also experimenting with market mechanisms, fiscal instruments, and regulatory interventions to finance climate change mitigation measures. One such source is the National Clean Environment Fund, which is funded by a tax on coal, currently at $3.2/ton. This had levied a total of $2.7 billion by 2015. The National Institute for Transforming India estimates, however, that moderate low-carbon mitigation activities will cost more than $800 billion by 2030 (2011 prices). Substantial international support will be required in financing, technology transfer, skill development, and capacity-building.

10.3.5 MALDIVES

HDI (Rank)	GNI/Capita	CRI (Rank)	EVI (Rank)	HAI (Rank)	Conditional
0.719 (104)	$12,549	103.3 (118)	50.9 (21)	91.4 (49)	True

GHG Target: 10% (unconditional) up to 24% (conditional) reduction in GHG emissions for the year 2030 compared to the BAU scenario. (Source: Ministry of Environment and Energy, Government of Maldives (2015).)

The Maldives, located in the Indian Ocean, west of the southern tip of India, is an island nation composed of almost 1,200 islands. The country's population of approximately 347,000 (as of 2014) is distributed across about 200 islands. According to the Ministry of Environment and Energy, fisheries and tourism are the most important economic sectors, both of which are sensitive to the impacts of climate change. Despite this sensitivity, the Climate Risk Index, which has only been measured since 2016, is squarely in the least vulnerable two-thirds of all countries. The Maldives is the second least vulnerable (followed by Bhutan) nation in SAARC. Though the country has the highest GNI per capita in the region, it also has the highest UN economic vulnerability, ranked 21st in the world.

Being such a small country with few energy-related natural resources, the Maldives faces a unique set of energy challenges. The country's energy demand is met almost entirely by imported fossil fuels; approximately 0.1% of the 2011 energy generated in the Maldives was from solar PV. Emissions in 2011 were around 1 million tons CO_2 equivalent, a mere 0.003% of global emissions, most of which are carbon dioxide and methane. Under the BAU scenario, the Maldives expects GHG emissions to increase more than three-fold, to 3.3 million tons of CO_2e by 2030. The INDC documents an unconditional commitment to reduce this by 10% to 2.97 million tons of CO_2e. Conditional upon sufficient external support, the country expects to reduce this by a further 15%, with 2.5 million tons of CO_2e forecast GHG emissions in 2030. These three projection curves are shown in Figure 10.9.

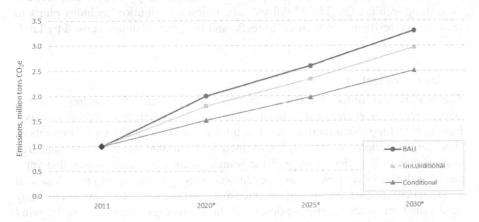

FIGURE 10.9 GHG emissions forecasts under different scenarios. (Source: Ministry of Environment and Energy, Government of Maldives (2015).)

Successful implementation of climate change adaptation and mitigation measures will rely heavily upon international support: adequate and predictable financial resources, transfer of environmentally-sound technologies, and capacity-building support. While mitigation is a key concern, adaptation is the primary focus.

10.3.5.1   Adaptation

Adaptation plans for the Maldives fall into nine strategic focus areas:

* enhancing food security
* infrastructure resilience
* public health
* enhancing water security
* coastal protection
* safeguarding the coral reef and its biodiversity
* tourism
* fisheries
* early warning system and systematic observation

Perhaps the most important focal topic to the Maldives is food security. With limited water resources, poor soil, and land scarcity, domestic agriculture and food production is extremely minimal. The majority of food, even staples, is imported and the country lacks adequate storage and distribution. Adaptation plans to enhance food security include strengthening climate risk insurance mechanisms for farmers, and the establishment of robust strategic food storage and distribution facilities. Safeguarding the coral reef and reef biodiversity is also of special importance not just for the Maldives, but for everyone. Coral reefs are an important source of local ecotourism income, and they also support fisheries. Furthermore, coral reefs fill several important roles in pelagic and coastal ecology. Considering the bigger picture, coral reefs are an important carbon sink, through the production of calcium carbonate ($CaCO_3$). According to Kinsey and Hopley (1991), coral reefs globally sequester 111 million tons of carbon annually — approximately 2% of anthropogenic CO_2. The Maldives' adaptation contribution includes plans to increase the resilience of their coral reefs, and to reduce damage caused by land-based pollution.

10.3.5.2   Mitigation

The Maldives considers the development and deployment of indigenous renewable sources to meet energy demand and ensure energy security strictly imperative. Indeed, switching to alternative energy options is the main focal area for mitigation. However, the options are severely constrained by the limited and disjointed land area available, as well as the isolation of the islands. Solar irradiance is available year-round, and may be a viable option for renewable energy, with appropriate financial and technical support. Because of the extensive coastal areas, tidal-wave energy may be a viable renewable energy option once the technology matures. Finally, wind resources in the Maldives are considered to be quite limited, due to the low lying and flat nature of the country.

10.3.6 NEPAL

HDI (Rank)	GNI/ Capita	CRI (Rank)	EVI (Rank)	HAI (Rank)	Conditional
0.579 (147)	$2,748	29.7 (20)	28.4 (95)	71.2 (95)	N/A

GHG Target: Not applicable (N/A). (Source: Government of Nepal, Ministry of Population and Environment (2016).)

Nepal is classified by the United Nations as an LDC, and has been in every triennial evaluation since 2000. Nepal's Human Development Index ranking is in the lowest quartile globally; the country's GNI per capita is the second lowest in the region, and in the lowest 15th percentile globally. Along with low income and depressed human development, Nepal's most recent Climate Risk Index ranking is just below the top decile. In the 2019 (2017 data) CRI evaluation, the country was ranked 4th globally. Climate change vulnerability in Nepal is predominantly related to glacial melting.

Between 1977 and 2010, estimated Himalayan ice reserves have decreased by nearly 30%. The number of glacial lakes has increased by 11%, and glaciers have receded at an average annual rate of 38 km². All of this is tied to higher risks of glacial lake outburst flooding. In addition, Nepal has experienced extreme weather events such as floods, landslides, and droughts more frequently. Higher temperatures and lower rainfall have reduced the amount of hydropower, which accounted for more than 99% of energy generated domestically, and approximately 65% of consumption in 2017 (International Energy Agency, n.d.). Under the scenario of lower projected hydropower generation, the additional energy capacity needed for future demand has been estimated at 2,800 MW by 2050, costing over $2.5 billion. Figure 10.10 shows quite clearly how the percentage of domestic electricity demand met by hydropower has decreased from 73% in 2014 to 64% — a nearly 10% decrease — while total demand increased by approximately 40%. The vast majority of this shortfall was met by imports.

A 2013 study entitled "Economic Impact Assessment of Climate Change in Key Sectors" has suggested that the costs of climate change thus far are approximately

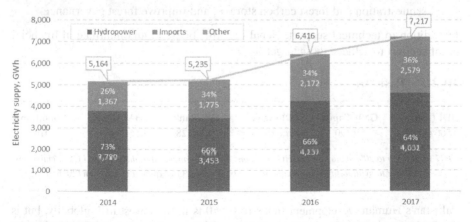

FIGURE 10.10 Nepal electricity supply by source. (Source: International Energy Agency (n.d.).)

1.5–2% of Nepal's GDP. This is projected to increase to up to 3% of GDP by the middle of this century.

Nepal has initiated several activities to mitigate the effects of climate change and develop resilience, and to reduce the impacts of climate change on its people, property, and natural resources. Broadly, these activities can be categorized as focused on institutions, policies/strategies/frameworks, adaptation actions, knowledge management, and mitigation actions. The country's INDC identifies several targets, which include:

- Nepal plans to formulate the Low Carbon Economic Development Strategy that will envision the country's future plan to promote economic development through low carbon emission with particular focus on: (i) energy, (ii) agriculture and livestock, (iii) forests, (iv) industry, (v) human settlements and wastes, (vi) transport, and (vii) commercial sectors.
- By 2050, Nepal will achieve 80% electrification through renewable energy sources having an appropriate energy mix. Nepal will also reduce its dependency on fossil fuels by 50%.
- Nepal aims to achieve the following targets under the National Rural Renewable Energy Programme, reducing its dependency on biomass and making it more efficient:
  - Mini and Micro Hydropower: 25 MW
  - Solar Home System: 600,000 systems
  - Institutional Solar Power Systems (solar PV and solar pumping systems): 1,500 systems
  - Improved Water Mill: 4,000 number
  - Improved Cooking Stoves: 475,000 stoves
  - Biogas: 130,000 household systems, 1,000 institutional and 200 community biogas plants
- Nepal will develop its electrical (hydropowered) rail network by 2040 to support mass transportation of goods and public commuting.
- Nepal will maintain 40% of the total area of the country under forest cover, and forest productivity and products will be increased through sustainable management of forests. Emphasis will equally be given to enhance carbon sequestration and forest carbon storage, and improve forest governance.

In addition to technical support, Nepal requires bilateral and multilateral financial grant support to achieve these targets.

10.3.7 PAKISTAN

HDI (Rank)	GNI/ Capita	CRI (Rank)	EVI (Rank)	HAI (Rank)	Conditional
0.560 (152)	$5,190	87.8 (100)	21.9 (128)	56.7 (114)	True

GHG Target: Up to 20% reduction in GHG emissions by 2030 compared to the projected GHG emissions, subject to availability of international grants. (Source: Government of Pakistan (2016).)

Pakistan's Human Development Index of 0.560 is in the lowest fifth globally, but is the 2nd lowest among South Asian countries. The country's average CRI during the

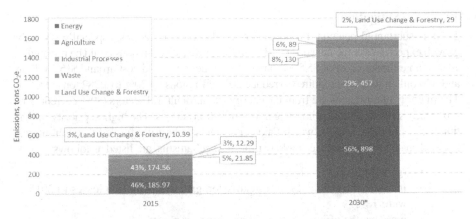

FIGURE 10.11   GHG emissions forecasts by sector. (Source: Government of Pakistan (2016).)

decade ending 2015 was ranked 7th highest in the world, but this plummeted in 2016. Indeed, as documented in the INDC, studies by the National Disaster Management Authority identified an average annual loss from extreme climate events of $4 billion between 1994 and 2013. A 2015 heatwave in Karachi cost more than 1,200 lives. Annual flooding between 2010 and 2014 caused over $18 billion in losses, affected more than 38 million people, and destroyed 10.6 million acres of crops. Despite financial constraints, the government reacted decisively; between 5.8% and 7.6% of the 2015 public budget was allocated for climate change-related expenditures. Currently, Pakistan's climate risk ranking has dropped to 100, putting it below the global median.

The majority of Pakistan's emissions are from agriculture and energy — approximately 90% of all GHG emissions. Ambitious governmental plans to spark economic growth through large-scale investments in energy, communication, and industrial infrastructure are expected to steadily increase emissions. This growth will likely coincide with an increased share of emissions from other sectors, such as industry and waste. As shown in Figure 10.11, emissions are expected to increase nearly four-fold between 2015 and 2030.

10.3.7.1 Adaptation

Pakistan's financial needs for climate change adaptation are estimated to fall between $7 billion and $14 billion annually; approximately $2 billion to $4 billion of this is just for adapting to flood disasters. The government of Pakistan is developing a National Adaptation Plan to create the framework for planning and implementing adaptation activities. Activities will be funded by balancing domestic contributions with the necessary international support. In addition to the National Adaptation Plan, Pakistan will develop and adopt sub-national adaptation plans. Medium- to long-term actions which will be pursued include:

- Improving the irrigation system through actions such as lining of canals and irrigation channels
- enhancing water resource management through integrated watershed management, water conservation, and optimization of water resource allocation
- strengthening risk management for the agriculture sector

10.3.7.2 Mitigation

There is substantial potential for mitigation in all economic sectors, especially given the growth expected over the next decade. The higher projected emissions could be reduced by 10% at an estimated cost of $5.5 billion; a 15% reduction would cost around $15.6 billion, and $40 billion would be required to reduce 2030 emissions by 20% to 1,282 tons of CO_2e. The majority of emissions are from the energy and agricultural sectors (89% current, 85% projected), so these are the primary foci for mitigation actions. The highest priority mitigation actions in these sectors, which Pakistan expects to be fully funded through international climate financing and technology transfer mechanisms, are listed as follows:

- Energy supply
  - increase the efficiency of transmission grid (transmission losses in 2015 were 18%)
  - increase the efficiency of coal combustion plants
  - implement utility-scale distributed grid-connected solar, wind, and hydroelectricity
- Energy demand
  - increase the efficiency of irrigation pumps and motors
  - replace incandescent bulbs with LEDs
  - increase the efficiency of cooking stoves
- Agriculture
  - improve irrigation and water management
  - manage water in rice cultivation to control the release of methane from soils and introduce low water-dependent rice varieties
  - implement agroforestry practices by planting multipurpose and fast-growing tree species

10.3.8 Sri Lanka

HDI (Rank)	GNI / Capita	CRI (Rank)	EVI (Rank)	HAI (Rank)	Conditional
0.780 (71)	$11,611	19.0 (6)	25.0 (114)	91.9 (47)	True

GHG Target: 4% (unconditional) to 20% (conditional) reduction in GHG emissions from the energy sector, and 3% (unconditional) to 10% (conditional) from other sectors (transport, industry, forests and waste) by 2030 compared to the BAU scenario. (Source: Ministry of Mahaweli Development and Environment (2016).)

Among the eight SAARC member states, the two island nations have the highest gross national income per capita and Human Development Index: Maldives and Sri Lanka. Sri Lanka's HDI of 0.780 barely beat the Maldives (0.719) for the top in the region, while Sri Lanka's GNI per capita is almost $1,000 less. Globally, Sri Lanka has the unique distinction of having the global median GNI per capita. A low-lying island in the midst of the Indian Ocean, Sri Lanka is particularly susceptible to variations in sea level, which disproportionately hit important economic sectors of agriculture, fisheries, and tourism.

Unique among SAARC countries, Sri Lanka's INDC covers three main topics: Adaptation, Mitigation, and Loss and Damage. Effective implementation of the actions proscribed under these aegises in the INDC is dependent upon three conditions. Firstly, sufficient external financial support is a crucial factor. Sri Lankan public funds have been, and will be, expended in support of the INDC targets. However, achieving the full conditional targets will rely on international funding. Secondly, technology plays an important role in both adaptation and mitigation activities. Sri Lanka requires access to relevant technologies, without hindering the country's socioeconomic development. Thirdly, support in capacity building is essential, particularly in developing institutional mechanisms.

10.3.8.1   Adaptation

Sri Lanka has identified eight sectors that are the most vulnerable to the effects of climate change. Selected contributions for these sectors — Health, Food Security (Agriculture, Livestock, and Fisheries), Water, Irrigation, Coastal and Marine, Biodiversity, Urban Infrastructure and Human Settlement, Tourism and Recreation — are detailed in Table 10.8. When possible, activities which align synergistically with mitigation measures will be given higher priority, so as to capitalize on the mitigation co-benefits.

TABLE 10.9
Key Adaptation Measures in Sri Lanka

Sector	Activities
Health	Establishment of clinical waste disposal systems in all hospitals
	Control of vector-borne and rodent-borne diseases
Food Security	Promotion/introduction/development of Integrated Pest Management practices (agriculture)
	Introduction of improved feeding practices (livestock)
	Establishment of fish barricade devices for each perennial reservoir (fisheries)
Water	Establishment of erect sandbags across the river during the drought season to prevent saline water intrusion wherever intakes are subjected to saline water intrusion
	Implementation of new water supply projects and schemes in water scarcity areas
Irrigation	Restoration and rehabilitation of all abandoned tanks and irrigation canals
	Establishment of water flow and sediment loads monitoring system in selected streams in the Central Highlands
Coastal/ Marine	Establishment of accurate sea level rise forecasting system
	Restoration, conservation, and management of coral, seagrass, mangroves, and sand dunes in sensitive areas
Biodiversity	Restoration of degraded areas inside and outside the protected area network
	Increasing connectivity through corridors, landscape/matrix improvement and management
Urban Infrastructure	Mainstreaming climate reliance in physical and urban planning and incorporating them into development projects
	Promotion of climate resilience building designing and alternative materials for construction
Tourism/ Recreation	Adjustment of tourism and recreation industry to altered conditions of the destinations
	Increase of preparedness of tourism and recreation operation to extreme weather conditions

10.3.8.2 Mitigation

Sri Lanka's INDC predominantly focuses on mitigation strategies to reduce GHG emissions from five sectors: electricity generation, transport, industry, forestry, and waste. Conditional contributions are expected to reduce emissions from the energy sector by 20% while reducing emissions from these other four sectors by half that amount. Unconditional reduction targets are 4% and 3%, respectively. Selected planned mitigation contribution actions are:

- establishment of large-scale wind power farms of 514 MW, and 115 MW of solar power plants
- establishment of energy-efficient and environmentally sustainable transport systems by 2030
- introduction of low-emission vehicles (electric and hybrid) into the system
- continuation of fuel switching to biomass in industries
- increase of forest cover of Sri Lanka from 29% to 32% by 2030

10.3.8.3 Loss and Damage

As climate change increases the frequency and severity of natural disasters, the risks of social, economic, and environmental vulnerabilities are also increasing. Between 2007 and 2011, the National Disaster Relief Services Centre records indicate that nearly $13 million of Sri Lankan public funds were spent on relief for climate change-related extreme weather events. This is likely a lower bound, as the accounting did not consider infrastructure damage. The Ministry of Disaster Management has estimated that the 2010 floods in the western and southern provinces of the island cost the public over US$35 million dollars. Sri Lanka's contributions for loss and damage include:

- improvement of forecasting capabilities — at all-time scales
- improvement of weather forecasting capabilities — extended range forecasting (longer period) and seasonal forecasting
- strengthening existing national mechanisms to recover losses and damages to the maximum possible extent

10.4 CHALLENGES

The South Asian region has always been an energy-deficient region, though not due to a lack of natural resources. Instead, this deficiency can be attributed to the inefficient use of resources, large population, primitive use of technology, lack of financial investment, poor management of distribution networks, slow structural improvement, and subsidies. There has been progress in the region in technological development, but development is slow due to restricted financial investments. Smart electricity systems are being implemented in the region, but mostly in urban areas. Energy access remains a challenge for the region, and the biggest hindrance to economic development and social well-being of the society.

People in this region have long struggled for basic human needs — food, water, shelter, and security; now more than ever, energy access is required to meet these needs. In general, the population is sensitive and efficient by nature in using

ENERGY SECURITY
Reflects a nation's capacity to meet current
and future energy demand reliably, withstand
and bounce back swiftly from system shocks
with minimal disruption to supplies.

ENERGY EQUITY
Assesses a country's ability to provide universal
access to affordable, fairly priced and abundant
energy for domestic and commercial use.

**ENVIRONMENTAL SUSTAINABILITY
OF ENERGY SYSTEMS**
Represents the transition of a country's energy
system towards mitigating and avoiding
potential environmental harm and climate
change impacts.

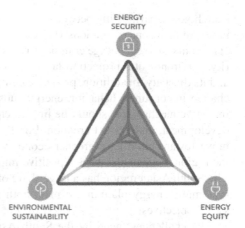

FIGURE 10.12 Energy Trilemma. (Source: World Energy Council & Oliver Wyman, 2019; Used by permission of the World Energy Council.)

resources at the individual level; whether it is electricity or fuel, people are sensitive and avoid wastage. This can be traced to the region's psychological profile, and can be seen in choices made: from automobile purchases, to the use of space in construction, to financial savings, to space exploration programs. However, conservation does not always mean being efficient. Energy efficiency is a challenge, not from the demand side but from the supply side. Convincing the population to change their behavior or make investments for energy efficiency at the residential, industrial, or commercial level can occur quite easily if coupled with regulatory requirements and with demonstration of the financial savings. Most importantly, though, energy efficiency needs to be implemented on the supply side. The structural improvement of transmission networks, grid management, transformers, and electric utilities are all much needed. In most countries of the region, the network system is primitive and frequently fails, resulting in transmission losses and interrupted supply. This is especially true during peak load, or summer months when demand loads are higher. Unfortunately, the untrustworthy energy supply system creates stress on the environmental systems by leading to deforestation to meet the population's basic needs of cooking, heating, and, water supply. The overuse of electricity for agriculture to mitigate the intermittent supply issue is also observed.

These challenges can be broadly summarized under the framework of the *Energy Trilemma,* shown in Figure 10.12:

"The World Energy Council's definition of energy sustainability is based on three core dimensions: Energy Security, Energy Equity, and Environmental Sustainability of Energy Systems. Balancing these three goals constitutes a "Trilemma" and balanced systems enable prosperity and competitiveness of individual countries" (World Energy Council & Oliver Wyman, 2019, p.13).

The conceptual framework is difficult to achieve, as there is a need for several checks and balances. It is a complex framework requiring integration between various stakeholders, partnership between public and private enterprises, and the implementation of stringent regulations (building codes, permits, energy efficiency ratings, carbon taxes) to encourage emissions reductions and energy supply access and security. Environmental protection laws are equally important to protect the biosphere and its diversity. In addition, proactive campaigns are needed to inform and promote change in consumer behavior, energy subsidies need to be reduced/abolished, and incentive mechanisms should be implemented to support innovation, technological development, and market creation. Energy transition requires effective planning to move toward building a greener economy that will not only help the countries in the region, but can also have a positive impact on mitigating global climate change. The South Asian region has a large share of the global population, and thus requires systematic energy planning to meet both short-term and long-term economic and social objectives.

The challenges faced by the South Asian countries can be resolved by applying the Energy Trilemma framework. In fact, balancing the Energy Trilemma on the three pillars of security, equity, and sustainability can both resolve the current challenges and provide multiple energy and non-energy benefits (Krarti et al., 2019) such as:

- job creation
- increased social well-being
- economic development
- environmental protection
- increased living standards

A top-down approach, keeping the three dimensions as the objectives, can assist SAARC member nations in resolving their energy challenges. These three dimensions are explored further in the remainder of this section.

10.4.1 ENERGY SECURITY

This section details four aspects of energy security which are particularly important to SAARC countries: security of supply, financial investment, technological advancement, and cross-border energy transmission.

- Security of Supply – Energy policies are important, due to the economy's reliance on energy. For energy-importing nations, energy security is imperative for both access and economic development. Conversely, energy policies can boost earnings from resource export for energy-exporting nations. A strong global network of energy trade interdependencies exists, which has led to collaboration on international climate agreements to reduce GHG emissions. Security of energy is required to provide continuous flow of energy catering to the requirements of industries, commercial entities, and residences.
- Financial Investment – Investments are necessary to change the current dynamics under which the government enjoys a monopoly of controlling energy supply and distribution networks. There should be a consideration to

open the market and call for investment in old and redundant infrastructure which is in dire need of maintenance. Private financial investments will provide a fiscal boost to the government and help in developing the market. This, in turn, benefits consumers with better service (and sometimes lower prices).

- Technological Advancement – Breaking the energy monopoly will give way for private entities to deploy new technologies and improve the current infrastructure. The common problem of transmission network failure can be overcome by implementing efficient technologies and systems that can be managed remotely.
- Cross-border Energy Transmission – Progress on the *SAARC Energy Ring*, first announced in 2004, has been slow and sporadic. More than a decade later, this may finally be the right time for the idea. The current situation is compelling and offers an opportunity for all the countries in the region to be connected and develop electricity trade within the region. Harnessing renewable energy, low-carbon technologies can simultaneously create a regional uninterrupted energy supply and provide further impetus to develop economically. A well-implemented integrated power grid in the South Asian region will deliver multiple benefits: affordability, reliability, and sustainability of electrical supply. It can also build the foundation for a future low-carbon electricity system. These benefits have been realized from power grid connections in other regional cooperatives. Examples include the European continent grid, Southern Africa Power Pool, and Central American Interconnection System.

10.4.2 Energy Equity

Five critical factors of energy equity are energy access, social well-being, subsidies, building codes, and tariffs. Their importance and relevance in South Asia is documented in this section.

- Energy Access – United Nations Sustainable Development Goals, SDG 7.1 (universal access to modern energy), and SDG 1.4 (access to basic services) are built on the assumption that energy is imperative for life, and a right of every person. In most South Asian countries, energy access is patchy. The urban centers mostly have access to electricity and other fuels, but intermittency is an issue. Many rural areas in these countries have limited access, possibly for a few hours a day, or none at all. These regions must rely on burning biomass and waste for cooking and heating. The lack of energy supply also places weight on the natural environment, leading to deforestation. Remote rural locations can be served by renewable and low-energy distributed generators, providing a long-term solution to reducing fossil fuel dependency.
- Social Well-being – Research has shown that people live healthier and are more productive if they live in energy-efficient homes. Providing energy for heating and cooling in well-insulated and energy-efficient buildings improves the well-being of the people. Labor productivity is equally important for the economic development of a country. Such non-energy benefits can have a wider economic impact on the society.

- Subsidies – Providing energy subsidies for industries or for people living below the poverty line has exhausted the resources of this region. Subsidizing the high proportion of the population living in poverty is a considerable financial burden on the system, more so when energy is imported. Subsidies can be replaced by adopting incentive mechanisms to promote efficient energy consumption. For example, developing an effective public transport system would help to remove the fuel subsidy. In addition, using renewable energy to power buildings and implementing Time of Use tariffs can ease pressure on the electricity grid during peak times. Mandatory installation of renewable technologies in commercial and public buildings, and roof-top solar water heaters on residential buildings has been a success in many countries, helping to reduce energy consumption.
- Building Codes – Mandating more stringent codes for new construction and retrofitting old buildings can help reduce energy waste, making buildings more self-sufficient. Buildings can become an energy sink and weigh on the society, but making better building codes mandatory can improve health and safety of the residents, as well as help provide continuous energy access.
- Tariffs – Feed-in tariffs have successfully changed the energy consumption patterns in Germany and many developed economies. Developing countries can equally benefit from opening the market and incentivizing people to install renewable energy technologies to meet their own energy needs, as well as sell the excess to the grid. This will help reduce energy imports.

10.4.3 Environmental Sustainability of Energy Systems

This section discusses four facets of the third dimension of the energy trilemma - environmental sustainability – that are of primary importance among South Asian countries: low carbon energy transition, competitive advantage, building resilience, and green growth and green economy.

- Low Carbon Energy Transition – Innovation to increase energy generation from clean and renewable sources and reduce energy demand through energy efficiency is imperative for the region. Decarbonizing its energy system, reducing waste through energy efficient technologies, and reducing dependency on fuel imports can significantly help the region overcome the energy and environmental hurdles currently being faced.
- Competitive Advantage – The South Asian countries have a substantial competitive advantage over many developed nations. They can use the learning curve and develop their economy based on sustainable foundations. Also, there are myriad opportunities to develop international cooperation through technological exchange. The mechanism of clean development is one of the tools that can provide a springboard for these developing countries to meet their dual goals of socioeconomic development and GHG emissions reduction.
- Building Resilience – There are many options for sustainable development (primarily energy conservation and GHG emission reduction), such as renewable energy, alternative energy, carbon capture and storage, green building innovations, smart cities, and sustainable transport. Further, the

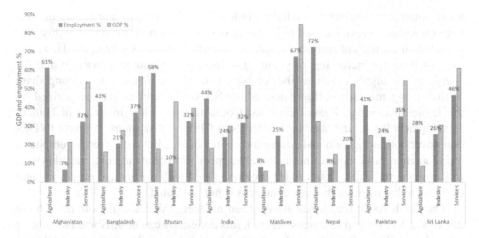

FIGURE 10.13 Value added and employment. (Source: International Labour Organization (2019) and United Nations Statistics Division (2019).)

interconnectivity between sectors and countries means that SAARC countries must work together to identify and adopt the jointly optimal approach to evolving a green economy that delivers prosperity through the growth of green industry, jobs, and earnings.

- Green Growth and Green Economy – Green growth places emphasis more neutrally on how energy can be used to drive economic growth, secure high-quality jobs, and increase public revenues. Green and sustainable economy concepts include: reducing carbon emissions, increasing energy efficiency, creating high-value jobs, and diversifying to a low energy-intensive economy. By developing the green economy agenda, the overall dependence on fossil fuels can be reduced by developing efficient transition pathways, resulting in improved economic and social welfare. Figure 10.13 shows the value creation from three economic sectors across South Asia: agriculture, industry, and services. It is of particular interest that, with the exception of the Maldives, all member states predominantly deploy labor in the agriculture sector, which contributes the least to GDP. A green economy perspective could pave the way for the sorely needed structural shift in these economies away by creating job opportunities at all levels. The green economy successfully links economic growth, reduces GHG emissions, creates jobs, and has the potential to develop a climate-resilient, resource-efficient, and socially inclusive economy by undertaking innovative business plans, financial initiatives, and pragmatic and cohesive policies.

10.5 CONCLUSION

Despite the vast land area covered by the eight SAARC member states—over 5 million km² — and the diversity of climate zones (Himalayan forest to desert to coastal reef), there are many commonalities regarding energy and the environment.

Most South Asian countries are lightly industrialized, with a strong economic reliance on agriculture. Indeed, all rely heavily on economic activity that is highly vulnerable to the effects of climate change: agriculture, fisheries, tourism, etc. However, most of them are increasing, or planning to increase, industrial activity. While the energy mix is highly disparate, the combustion of biofuels/waste has a strong share in the energy mix across the region. Many SAARC countries are categorized by the United Nations as Least Developed Countries, rated low in terms of Human Development Index, and rated high in terms of Climate Risk Index. All have INDCs that cover both mitigation and adaptation contributions, but with stronger emphasis on adaptation, and require international support to implement said contributions. In summary, the entire South Asian region is faced with energy and environmental challenges that threaten dire consequences if not mitigated.

The South Asian region has always been an energy-deficient region, though not due to a lack of natural resources. Instead, this deficiency can be attributed to, among other things, the inefficient use of resources, lack of financial investments, and poor management of infrastructure. There has been progress in the region in technological development, but development is slow due to limited financial resources. We have suggested several ways by which SAARC member countries can resolve their energy challenges, using the Energy Trilemma framework. Balancing the Energy Trilemma on the three pillars of security, equity, and sustainability can both resolve the current challenges and provide multiple energy and non-energy benefits, driving sustainable socioeconomic development.

REFERENCES

Alam, F., Saleque, K., Alam, Q., Mustary, I., Chowdhury, H., & Jazar, R. (2019). *Dependence on energy in South Asia and the need for a regional solution*. Energy Procedia, 160, 26–33. doi:10.1016/j.egypro.2019.02.114

Central Intelligence Agency. (n.d.). *The World Factbook*. Retrieved from https://www.cia.gov/library/publications/the-world-factbook/; https://www.indexmundi.com/map/?v=81&r=as&l=en

Dow, K., & Downing, T. (2011). *The Atlas of Climate Change*. Brighton, East Sussex: Earthscan.

Eckstein, D., Künzel, V., Schäfer, L., & Winges, M. (2019). *Global Climate Risk Index 2020: Who Suffers Most from Extreme Weather Events? Weather-Related Loss Events in 2018 and 1999 to 2018*. Retrieved from Germanwatch: https://germanwatch.org/en/17307

Eckstein, D., Hutfils, M., & Winges, M. (2018). *Global Climate Risk Index 2019: Who Suffers Most from Extreme Weather Events? Weather-Related Loss Events in 2017 and 1998 to 2017*. Retrieved from Germanwatch: https://germanwatch.org/en/16046

Eckstein, D., Künzel, V., & Schäfer, L. (2017). *Global Climate Risk Index 2018: Who Suffers Most from Extreme Weather Events? Weather-Related Loss Events in 2016 and 1997 to 2016*. Retrieved from Germanwatch: https://germanwatch.org/node/14987

Government of India. (2015). *India's Intended Nationally Determined Contribution: Working toward Climate Justice*. Retrieved from https://www4.unfccc.int/sites/submissions/indc/Submission%20Pages/submissions.aps

Government of Nepal, Ministry of Population and Environment. (2016). *Intended Nationally Determined Contributions (INDC)*. Retrieved from https://www4.unfccc.int/sites/submissions/indc/Submission%20Pages/submissions.apx

Government of Pakistan. (2016). *Pakistan's Intended Nationally Determined Contribution (PAK-INDC).* Retrieved from https://www4.unfccc.int/sites/submissions/indc/Submission%20Pages/submissions.aspx

Human Development Report Office. (2019). *Beyond Income, Beyond Averages, Beyond Today: Inequalities in Human Development in the 21st Century.* Retrieved from http://hdr.undp.org/en/content/2019-human-development-index-ranking

International Energy Agency. (n.d.). *Data Tables – Data & Statistics.* Retrieved from https://www.iea.org/data-and-statistics/data-tables

International Labour Organization. (2019). *Key Indicators of the Labour Market* (9th Edition). Retrieved from http://data.un.org/_Docs/SYB/PDFs/SYB62_200_201905_Employment.pdf

Islamic Republic of Afghanistan. (2015). *Intended Nationally Determined Contribution Submission to the United Nations Framework Convention on Climate Change.* Retrieved from https://www4.unfccc.int/sites/submissions/indc/Submission%20Pages/submissions.aspx

Kinsey, D., & Hopley, D. (1991). *The significance of coral reefs as global carbon sinks — response to greenhouse.* Global and Planetary Change, 3(4), 363–377. doi:10.1016/0921-8181(91)90117-f

Krarti, M., Dubey, K., & Howarth, N. (2019). *Energy productivity analysis framework for buildings: a case study of GCC region.* Energy, 167, 1251–1265. doi:10.1016/j.energy.2018.11.060

Kreft, S., Eckstein, D., & Melchior, I. (2016). *Global Climate Risk Index 2017: Who Suffers Most from Extreme Weather Events? Weather-Related Loss Events in 2015 and 1996 to 2015.* Retrieved from Germanwatch: https://germanwatch.org/12978

Ministry of Environment and Energy, Government of Maldives. (2015). *Maldives Intended Nationally Determined Contribution.* Retrieved from https://www4.unfccc.int/sites/submissions/indc/Submission%20Pages/submissions.aspx

Ministry of Environment and Forests, Government of the People's Republic of Bangladesh. (2015). *Intended Nationally Determined Contributions.* Retrieved from https://www4.unfccc.int/sites/submissions/indc/Submission%20Pages/submissions.apx

Ministry of Mahaweli Development and Environment. (2016). *Intended Nationally Determined Contribution.* Retrieved from https://www4.unfccc.int/sites/submissions/indc/Submission%20Pages/submissions.aspx

National Environment Commission, Royal Government of Bhutan. (2015). *Kingdom of Bhutan Intended Nationally Determined Contribution.* Retrieved from https://www4.unfccc.int/sites/submissions/indc/Submission%20Pages/submissions.aspx

Schwab, K. (2017). *The Global Competitiveness Report 2017-2018.* Retrieved from http://www3.weforum.org/docs/GCR2017-2018/05FullReport/TheGlobalCompetitivenessReport2017%E2%80%932018.pdf

United Nations Department of Economic and Social Affairs. (2018). *Criteria for Identification of LDCs.* Retrieved from https://www.un.org/development/desa/dpad/least-developed-country-category/ldc-data-retrieval.html

United Nations Department of Economic and Social Affairs, Population Division. (2019). *World Population Prospects 2019, Online Edition. Rev. 1.* Retrieved from https://population.un.org/wpp/Download/Standard/Population/

United Nations Statistics Division. (2019). *National Accounts Statistics: Analysis of Main Aggregates (AMA) Database.* Retrieved from http://data.un.org/_Docs/SYB/PDFs/SYB62_153_201906_Gross%20value%20added%20by%20kind%20of%20economic%20activity.pdf

World Energy Council, & Oliver Wyman. (2019). World Energy Trilemma Index, 2019. Retrieved from https://trilemma.worldenergy.org

11 Conclusion

Muhammad Asif

CONTENTS

11.1 Sustainable Energy Transition 253
11.2 Road to Energy and Environmental Security 254
    11.2.1 Success Stories 254
    11.2.2 Challenges 256
11.3 Navigating through the New Normal 259
11.4 Regional Cooperation 260
11.5 Concluding Remarks 262
References 263

11.1 SUSTAINABLE ENERGY TRANSITION

Human use of energy has evolved over the course of history. The availability of refined and efficient energy resources has played a key role in the advancement of societies especially since the industrial revolution of the 18th century. In the 21st century, the world is witnessing an unprecedented energy transition. Given the faced challenges—such as growth in energy demand, depleting fossil fuel reserves, increasing greenhouse gas emissions, volatile energy prices, and energy insecurity issues—the global energy scenario is experiencing a paradigm shift in terms of energy resources and their utilization. Historically, the world has seen major energy transitions in the form of switchovers from biomass and wood to coal in the 18th century, and from coal to oil and gas in the 20th century. The present energy transition, however, is much more dynamic and multidimensional compared to the earlier ones. With the main focus on sustainability, enormous changes are taking place in terms of energy resources and their utilization, technological and policy advancements, and socio-economic and political response. This energy transition is manifested by four major technological drives: decarbonization, energy conservation and management, distributed generation, and digitalization. These dimensions are at the heart of sustainable energy policies and frameworks across the world. The European Union's 20-20-20 directive is an interesting example in this respect which aimed to increase energy efficiency by 20%, reduce CO_2 emissions by 20%, and have renewables making up 20% of supplies by 2020.

Energy and environmental sustainability are at the heart of the ongoing energy transition. All of the four aforementioned dimensions of this energy transition are working toward a sustainable energy scenario. Decarbonization of the energy sector, for example, is being spearheaded by solutions such as renewable and low-carbon technologies, electric mobility, carbon capture and storage, and hydrogen and fuel cells. Similarly, energy conservation and management are pivotal to energy and environmental sustainability. Energy efficiency is recognized to be the easiest and the most cost-effective solution to growing energy needs since a unit of energy saved is better than a unit generated.

South Asian countries (SACs) need to be abreast of this energy transition. While some of the emerging solutions such as hydrogen and fuel cells, and digitalization technologies are yet to see the due levels of technical and economic maturity, there are a number of avenues that can be actively pursued. Renewable energy and energy efficiency, for example, are imperative to be adopted to address SACs' energy and environmental challenges, especially when they are relying on fossil fuels, a major proportion of which is imported. It is also important for the region to maximize supplies from indigenous resources to reduce dependency on energy imports.

11.2 ROAD TO ENERGY AND ENVIRONMENTAL SECURITY

In recent years the world has made significant progress in improving the access to energy. Since 2010, for example, the global population lacking access to electricity has dropped from 1.2 billion to below 800 million. This progress has been significantly helped by the off-grid renewable solutions, especially solar PV systems. Despite these advances, global efforts are not up to mark to meet SDG7 by 2030. In a Business As Usual (BAU) Scenario, in 2030, over 600 million people will still lack access to electricity, and 2.3 billion people will be relying on inefficient and crude biomass fuels and technologies (WB, 2020). South Asian countries have also made considerable progress. In recent years, the electrification of rural and remote communities, for example, is an area where significant improvements have been made. Governments have launched major grid extension initiatives. Off-grid renewable energy programs both on the part of public and private sector entities have also played a major role in providing electricity to communities lacking grid connectivity. The combined power generation capacity of the three largest countries in the region—India, Pakistan, and Bangladesh—grew from less than 200 gigawatts (GW) in 2007 to over 400 GW at the end of 2019. Over this period, the proportion of households having access to electricity in these countries has increased from less than 70% to over 85%.

11.2.1 SUCCESS STORIES

South Asian countries have seen a number of small-scale renewable energy programs initiated to provide modern energy services to rural and remote communities. Some of these programs have been quite successful such as Grameen Shakti in Bangladesh, Energy Services Delivery Project in Sri Lanka, and Renewable Energy for Rural Livelihood program in Nepal. These programs—mainly dealing with technologies such as solar home systems, biogas systems, hydropower, and improved cooking stoves—have partnered with various local and international entities including the World Bank (WB) and United Nations Development Program (UNDP). These energy initiatives have adopted different working models including microcredit, government subsidies, and communal investments. The micro-credit based model has, however, experienced the most success.

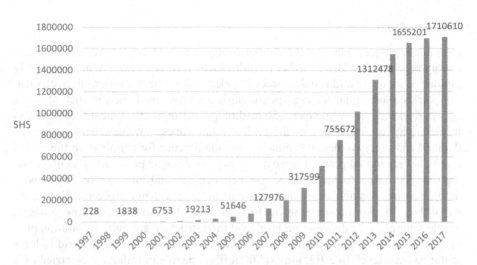

FIGURE 11.1 Annual growth of solar home systems installed by Grameen Shakti.

Grameen Shakti (GS) is a major success story in this respect. GS is one of the world's largest micro-generation renewable energy programs that started in 1996 in Bangladesh to provide affordable and environmentally friendly energy to remote communities in the country. Emphasis was placed to address the needs of the people lacking access to the grid (Asif and Barua, 2011). GS started by employing solar photovoltaic (PV) technology to install solar home systems (SHSs). Having realized the importance of electricity in the socio-economic well-being of people, GS started the SHS program in 1996. The success of SHSs is evident from the fact that the cumulative number of the installed SHSs increased from 228 in 1997 to over 1.7 million by the end of 2017 as shown in Figure 11.1. In 2005, GS launched a biogas systems program and has so far installed over 33,000 units. Similarly, in 2006 it introduced improved cooking stoves (ICSs) program and has thus far installed nearly 1 million systems. According to the Infrastructure Development Company Limited (IDCOL), a coordination body on small-scale renewable energy systems in Bangladesh, over 4.1 million SHSs have been installed in the remote areas of the country where electrification through grid expansion is challenging and costly. Around 18 million people are benefiting from the SHSs. IDCOL has a target to finance 6 million SHSs by 2021 with an estimated installed capacity of 220 MW (IDCOL, 2020).

Another success story is India's progress in modern renewables, i.e., wind power and solar energy. India has made significant accomplishments not only in terms of installed capacity, but also the technological base for wind power and solar energy. With a respective installed capacity figures of 37 GW, 33 GW, and 85 GW, India stands 4th, 6th, and 5th in wind power, solar energy, and renewable energy deployment in the world. It has set a target of having 175 GW of renewables by the year 2022 including 100 GW of solar energy, 60 GW of wind energy, 10 GW of biomass and bagasse, and 5 GW of small hydropower. Of this, as of 2019, 84.7 GW of projects have been installed (MNRE, 2020).

11.2.2 CHALLENGES

Despite the progress made on the electrification front, SACs still have a catalogue of energy problems to address. To begin with, electrification alone does not essentially guarantee adequate and reliable access to electricity. The definition of electrification varies a lot. India, for example, regards a village electrified if 10% of it has access to electricity. While over 350 million people in South Asia still lack access to electricity, the reliability of the grid in many cases is weak. There are severe demand and supply issues. It is quite a common practice for countries in the region to employ scheduled power blackouts or load-shedding even in urban centers and metropolitan cities to manage the gap between demand and supply. Issues with aging transmission and distribution systems are also a major problem. The fragility of the grid also transpires in the form of power breakdowns and low voltages. While failures at the distribution level grid infrastructure are quite a common phenomenon, in recent years both India and Pakistan have faced multiple grid failures at the national level. In 2012, almost 700 million people in India lost electricity for two days as a result of a power breakdown. Similarly, in 2015 almost 80% of the population in Pakistan lost power for several hours as a result of problems in a major transmission line. Pakistan experienced another nationwide power failure in 2018. Not surprisingly, South Asia has the highest number of power outages as compared to other developing regions of the world including Sub-Saharan Africa and Latin America as shown in Figure 11.2 (Zhang, 2019). India, Pakistan, and Bangladesh, the three biggest nations in the region, have very low standing in terms of grid reliability with respective positions of 108, 99, and 68 out of 140 countries in the world as shown in Table 11.1 (Schwab, 2019). While access to electricity helps improve the economic well-being of societies, grid reliability is also helpful in this respect as shown in Figure 11.3.

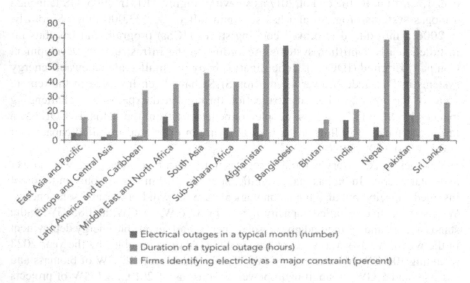

FIGURES 11.2 Reliability of grid in South Asia compared to other regions.

TABLE 11.1
Global Ranking of South Asian Countries in Grid Reliability

Indicator	Position in 140 Countries				
	India	Pakistan	Bangladesh	Nepal	Sri Lanka
Electrification access	105	111	108	101	2
Electric supply quality	108	99	68	119	39
Energy efficiency regulation	33	70	82	92	59
Renewable energy regulation	3	58	81	85	57
Exposure to unsafe drinking water	106	125	136	122	93
Reliability of water supply	96	106	115	125	83
Environment-related treaties in force	17	69	95	126	79

While access to reliable electricity is still a major issue for a substantial proportion of the regional population base, around 65% of the people in South Asia are
relying on crude biomass fuels such as wood, animal dung, and crop waste to meet
their cooking and heating requirements. In India alone over 700 million people use
biomass, while in Nepal it is the backbone of the national energy mix, making up
around 80% in total primary energy supplies.

South Asian Countries are also heavily suffering from energy sector inefficiencies.
There are wide-ranging leaks and losses across the energy sector, the economic impact
of which is colossal. According to the World Bank, the economic cost of the power sector distortion in India, Bangladesh, and Pakistan is 4–7% of the national GDP figures.
Distortions in India's power sector, for example, cost the national economy equivalent
to US$86 billion in the year 2016. Similarly, Pakistan's power sector in 2015 cost the
national economy around US$18 billion for the same reason (Zhang, 2019).

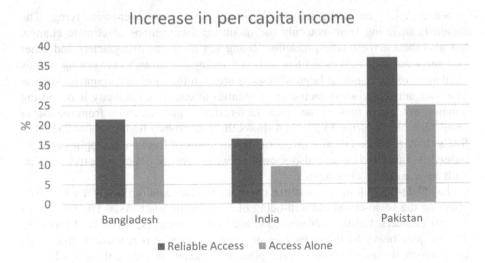

FIGURE 11.3 Role of access to electricity in improving income.

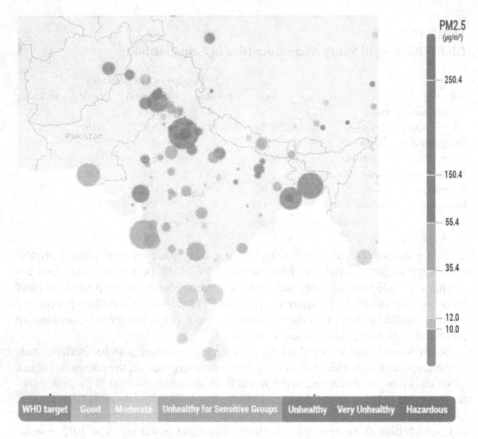

| WHO target | Good | Moderate | Unhealthy for Sensitive Groups | Unhealthy | Very Unhealthy | Hazardous |

FIGURE 11.4 Concentration levels of PM2.5 in SACs.

South Asia's environmental problems are severe and ever-intensifying. The region is suffering from not only the mounting implications of climate change, but also the microclimatic problems. Rising sea level, melting glaciers and other extreme weather-related calamities such as flooding and droughts are posing serious challenges to the region at large with some areas in the low-lying countries such as Maldives and Bangladesh facing an existential threat. Water scarcity is becoming a major problem in most of the SACs. Deforestation and emissions from the use of fossil fuels are on a rise. Figure 11.4 shows the dangerously high levels of PM2.5 in SACs (IQAir, 2019). The alarming situation in terms of freshwater supplies is highlighted in Table 11.1, which also shows the weak level of policy initiatives to deal with environmental challenges.

The wide-ranging micro- and macro-level environmental problems are seriously affecting the socio-economic well-being of societies in South Asia. The impacts of natural disasters, health problems, food and water shortages, and loss of biodiversity are quite heavy for the economies of these countries. It is noteworthy that while environmental concerns hardly find a place in national narratives, there is a lack of action on the part of governments especially in terms of policy development and implementation. Trends such as growing population and urbanization, increasing consumption of materials and resources, and surging levels of energy consumption and greenhouse gas (GHG) emissions are only exacerbating the environmental scenario.

11.3 NAVIGATING THROUGH THE NEW NORMAL

The COVID-19 pandemic has driven the world into uncharted territories. The pandemic is set to have monumental and far-reaching impacts on the human race and the planet. By the end of October 2020, there were over 51 million infected cases with more than 1.29 million deaths worldwide (JH, 2020). COVID-19 has ushered in unprecedented developments. Most of the countries across the world ordered their people to stay home. Airports and public transport were shut down. Borders, even within the European Union, were sealed. Social distancing was declared mandatory even where interaction was unavoidable. The economic and industrial wheel of the world appeared to have stalled. The socio-economic cost of the crisis, from local to global, is unparalleled. The consequent economic recession is being tipped to be the worst since World War II. In the absence of vaccination, estimates about the longevity of the COVID-19 crisis and the true scale of its fallout are quite murky. As scientific understanding of the make-up and dynamics of the virus is evolving, so are the projections about the post-COVID-19 world. While the best case scenarios suggest the vaccination to be available by the end of 2020, others indicate it may take longer. There are also the worst-case scenarios hinting at the possibility of never finding a vaccination, in which case societies may have to learn to live with the virus the hard way.

COVID-19 has arguably changed the world forever. The scope and intensity of the impacts would greatly depend upon how quickly and effectively vaccination is made available. Even if a cure is available by the end of 2020, the psychological impact, the fear factor, and uncertainties around it will remain there for a considerable period of time. There will be many things that will not return to the pre-COVID-19 era. The world has to live with a new normal, no matter how inconvenient it will be.

The energy sector is also feeling the impact of COVID-19. Oil prices have plunged due to the diminished demand with the United States recording negative prices in the latter half of April 2020. The energy sector is also experiencing supply cuts, disrupted trade flows, mounting debts, and bankruptcies. The full and long-term impact of the pandemic is hard to predict. Some of the short-term challenges are maintaining grid and networks fully functional, and navigating through low price and demand reduction scenarios. The Energy Information Administration (EIA) estimates that in the second quarter of 2020 the global demand for petroleum and liquid fuels will average 83.8 million barrels per day (b/d), compared to 100.4 million b/d at the same time last year. EIA forecasts that consumption for the full 2020 will average 92.5 million b/d compared to 100.8 million b/d from 2019. EIA expects the consumption to rise to 99.7 million b/d in 2021(EIA, 2020). Other challenges of the COVID-19 will include managing curbed revenue, liquidity shortage, and mounting debt obligations. Another implication of the pandemic on the energy sector is the disruption in the supply chain, especially around renewables. China—the biggest manufacturer of renewable energy components especially when it comes to solar technologies and wind turbines—being the epicenter of the COVID-19, saw its manufacturing base almost halted for several months. China has restarted production and manufacturing but with limited capacity. The delays in the supply chain could result in an increase in prices for renewable materials and components in the short term. The International Renewable Energy Agency (IRENA) is concerned that the pandemic may have an impact on the sustainable energy transition (IRENA, 2020). According to the International Hydropower Association (IHA), the prevalent uncertainty combined with liquidity shortages has put financing and refinancing of many hydropower projects at risk. New and upgrade projects are also being halted, contributing to a fall in confidence regarding future investments and operations (IHA, 2020). PricewaterhouseCoopers (PWC) has surveyed with industry leaders to determine their major concerns about the impact of COVID-19 on the energy sector. 71% of the respondents expressed their main concern as the financial impact towards the operation of current projects, future of the projects in pipeline, and liquidity and capital resources. Other major concerns recorded are potential global recession; the effects on workforce/reduction in productivity; decrease in consumer confidence reducing consumption; supply chain disruptions; difficulties with funding; not having enough information to make good decisions; and impacts on tax, trade, or immigration (PWC, 2020).

COVID-19 is set to have major implications for South Asia as well. Besides the loss of lives, the financial fallout is going to be colossal for the developing and underdeveloped economies of these nations. There will be micro- and macro-level implications on many fronts. According to the UN Labor Agency, around half of the global workforce could see their livelihood destroyed (UN, 2020a). Hundreds of millions of South Asians are already facing the financial heat in the form of job losses, reduced working hours and pay slashes; many of whom are likely to be pushed below the poverty line. The situation is going to affect the governments' capacity to focus and invest in addressing energy and environmental issues. In the wake of financial difficulties people are facing, the energy security situation especially in terms of affordability will become more critical.

11.4 REGIONAL COOPERATION

An important area of which South Asian countries have struggled to realize the potential and importance is transnational and regional trade and cooperation in the field of energy. Despite the fact that SACs heavily rely on fossil fuel imports to satisfy their energy requirements, they have ignored the effective utilization of energy resources within their backyard.

South Asia has quite a rich and diverse set of energy resources including renewables and fossil fuels. In renewables, it has abundant potential for hydropower, solar energy, wind power, and biomass. Bhutan, India, Nepal, and Pakistan, for example, have huge hydropower resources through the rivers originating from the Himalayan basin. The total hydropower potential in the region is estimated to be around 350 GW. Such a vast resource, though seasonal in nature, if adequately utilized, can make a big difference in improving the energy, environmental and socio-economic conditions of the whole region. Hydropower, as it depends on water use, has unfortunately become a sensitive subject in South Asia's national and regional politics. India and Pakistan also have massive potential for wind power, with the former already having an installed capacity of over 37 GW. India has vast coal reserves that are playing an important role in the country's power generation mix. Pakistan also has significant but untapped coal reserves. There are also considerable gas resources, especially in Pakistan and Bangladesh. Pooling together this diverse set of resources through interconnected grid and supply networks can help South Asia in terms of tackling its energy woes. It can help address the gap between energy demand and supply besides improving its affordability. (Khadka, 2012).

South Asian countries have taken some initiatives to promote regional cooperation in the field of energy. SAARC Energy Centre, for example, was created in 2005, but it has been largely limited to training and similar capacity-building activities. There have been some energy transactions under the South Asia Subregional Economic Cooperation (SASEC), mainly in the form of electricity trade between India-Bangladesh, Bhutan-India, and India-Nepal, and trade in petroleum products amongst these countries. (Wijayatunga and Fernando, 2013). The capacity of electricity transfer from India to Bangladesh, India to Nepal, and Bhutan to India is through projects of 600 MW, 330 MW, and 1,400 MW, respectively. Estimates suggest that by 2050, through measures such as effective policy coordination, grid interconnection, market reforms, trade agreements, and regulation, SACs have the potential to upscale the electricity trade up to over 70 GW of projects (Panda 2016). Apart from electricity trade, SACs have enormous potential to cooperate around fossil fuels especially oil and gas. In the recent past, several major projects have been explored to establish this cooperation within the region and beyond, for example, the Iran-Pakistan-India (IPI) natural gas pipeline and the Turkmenistan-Afghanistan-Pakistan-India (TAPI) natural gas pipeline. Similarly, high-capacity electricity links have also been explored such as the Central Asia-South Asia (CASA 1000) power link. The China-Pakistan Economic Corridor (CPEC) as part of China's "One Road One Belt" plan also offers a glaring opportunity to the whole region to establish a strong partnership in the field of energy.

According to the Asian Development Bank (ADB), South Asian countries need to strengthen regional energy cooperation to benefit from economies of scale through a vibrant intra- and inter-regional energy trade structure. It identifies four major areas to focus on to enhance energy cooperation in the region: a regional power market; energy supply availability; energy trade infrastructure; and harmonized legal and regulatory frameworks (Wijayatunga and Fernando, 2013). The UN's Economic and Social Commission for Asia and the Pacific (ESCAP) has also recommended the regional energy cooperation through measures such as effective and coordinated utilization of unequally distributed energy resources, infrastructure development, enhanced use of renewable energy and promotion of energy-efficient measures, and creation of sub-regional energy markets (ESCAP, 2013).

South Asian countries also need to join hands in the fight against the common threat of climate change. Over half of the world's population categorized as "absolute poor" lives in South Asia, the livelihood of many of whom depends on climate-sensitive sectors such as agriculture, forestry, and fishing. Owing to its impacts on agriculture and water cycle, climate change is also affecting food and water security. There is the inevitable knock-on effect on the poverty alleviation and economic development efforts in the region (ADB, 2020). SACs can also cooperate to address other environmental and micro-climatic challenges. India and Pakistan, for example, over the last few years, are facing recurring smog crisis that seriously affects the daily routines of a significant proportion of population in the two countries for several weeks every year, besides having severe economic, health and safety implications. Instead of getting into a blame game, the two sides can better tackle the issue through coordinated efforts.

The lack of mutual trust and the widespread geopolitical adversaries are the major hurdles towards the regional cooperation in the field of energy and climate change. Primarily, it is not technical or economic issues that have handicapped regional cooperation, but lack of political will on the part of the South Asian governments. The countries in the region have not demonstrated the due level of pragmatism to resolve their long-standing conflicts, for which they are paying a big price. Geopolitical disputes have not only made peace and stability in the region hostages, but are also bleeding the economies of these developing countries.

11.5 CONCLUDING REMARKS

Almost a quarter of the world's population residing in South Asia is facing mounting energy and environmental challenges. In recent years South Asian countries (SACs) have made considerable progress on the energy front especially in terms of growth in power generation capacity and electrification of rural communities. Over 350 million people still lack access to electricity and over 1 billion people have to use crude biomass fuels to satisfy their cooking and heating requirements. Many of those having access to electricity and gas networks still face issues such as unreliable grid and expensive energy prices. The energy outlook of the SACs also faces challenges such as rapid growth in demand, heavy reliance on imported fossil fuels, and the associated environmental emissions. Similarly, the environmental outlook of South Asia faces daunting threats from global warming and climate change with weather-related calamities such as flooding, hurricanes, and droughts becoming recurring phenomena affecting a major proportion of the population in the region. While rising sea level is a major concern for low-lying areas across the SACs, Maldives and Bangladesh in particular face a mounting threat from the issue. In a business as usual scenario, SACs are likely not to be anywhere closer to satisfying the United Nation's Sustainable Development Goals (SDGs) pertinent to the energy and environmental security—especially SDG 6, SDG 7, and SDG 11–15.

The South Asian countries are heavily suffering from mismanagement in the energy sector. According to the World Bank estimates, the collective economic cost of the power sector distortions in Bangladesh, India, and Pakistan is of the order of US$115 billion per annum (Zhang, 2019). Considering the overall energy sector, the cost of leaks and losses across South Asia would be substantially higher. Consequently, the SACs have a very low ranking in the world when it comes to energy management and governance. To improve their energy and environmental outlook, SACs need to better manage their energy sector. A robustly managed energy sector would reduce losses and improve grid reliability and access. It would also lead to reduction in cost, improvement in revenues, and decline in environmental emissions. The consequent economic gain would also be huge. According to the World Bank, few other reforms could quickly yield economic gains of a similar magnitude as compared to the reforms in the energy sector (Zhang, 2019). The situation with the environmental scenario is not much different either. Effective governance and management are thus imperative for these countries to sustainably address their energy and environmental challenges.

South Asia's energy and environmental outlook is being significantly undermined by the geopolitical adversaries in the region. Decades-old unresolved territorial and water disputes in the region—especially the India-Pakistan and India-China confrontations, with a history of wars, and all three being nuclear-armed states—pose a serious threat to the stability of the whole region. The situation between India and Pakistan becomes volatile every now and then, bringing them on the brink of war. India and China are also experiencing heightened border tensions in recent years, with the two sides being engaged in a months-long deadly border faceoff over disputed territories. With a growing trend, in 2019, over US$80 billion were spent by the SACs on the defense budget, while housing almost half of the world population below the poverty line and with per capita electricity consumption being less than a quarter of the world average. With the mounting energy- and environmental challenges, SACs need to revamp their national priorities. They need to optimally channelize their resources and efforts toward sustainable development instead of stimulating geopolitical adversaries. The SACs have tremendous potential to cooperate in the field of energy and benefit from the diverse range of resources, from within the region and beyond, to address their energy and environmental challenges. It is high time for the SACs to contemplate a future based upon commonalities and cooperation instead of hostilities.

The COVID-19 is a wake-up call for the South Asian countries to reflect and demonstrate sanity. It is time to explore a more resilient lifestyle based on sustainable technologies and practices. Human interaction with nature and natural resources needs a reset. According to the UN Secretary-General, António Guteres, "The recovery must also respect the rights of future generations, enhancing climate action aiming at carbon neutrality by 2050 and protecting biodiversity. We will need to build back better." The United Nations Development Programme Administrator, Achim Steiner, emphasizes that, "As we work through response and recovery from the shocks of the pandemic, the SDGs need to be designed into the DNA of global recovery" (UN, 2020b). According to the Executive Director of the International Energy Agency, Dr. Fatih Birol, "Governments have a once-in-a-lifetime opportunity to reboot their economies and bring a wave of new employment opportunities while accelerating the shift to a more resilient and cleaner energy future" (IEA, 2020). As Winston Churchill once said, "Never let a good crisis go to waste", the South Asian countries should avail the "Build Back Better" opportunities presented by the pandemic to make a positive contribution to their energy and environmental outlook.

REFERENCES

ADB (2020). Climate Change in South Asia: Strong Response for Building a Sustainable Future, Asian Development Bank.

Asif M (2011). Energy Crisis in Pakistan: Origins, Challenges and Sustainable Solutions, Oxford University Press, Karachi.

Asif M and Barua D (2011). Salient features of the Grameen Shakti Renewable Energy Program, *Renewable & Sustainable Energy Reviews*, 15, 9, 5063–5067.

EP. (2008). Climate Change and Energy Package 20 20 20 by 2020, European Parliament, 11 December 2008, http://www.europarl.europa.eu/sides/getDoc.do?type=IM-PRESS &reference=20081208BRI43933&secondRef=ITEM-002-EN&language=EN

EIA (2020), Short-Term Energy Outlook, EIA, 9 June 2020, https://www.eia.gov/outlooks/steo/

ESCAP (2013). Regional Cooperation for Energy Access and Energy Security in South and South-West Asia: Prospects and Challenges, Economic and Social Commission for Asia and the Pacific, United Nations, February 2013.

JH (2020). Coronavirus Resource Center, Johns Hopkins University & Medicine, https:// coronavirus.jhu.edu/map.html

IEA (2020). Sustainable Recovery, International Energy Agency, https://www.iea.org/reports/ sustainable-recovery

IHA (2020). Hydropower associations unite to set Covid-19 recovery pathway, 18 May 2020, https://www.hydropower.org/news/hydropower-associations-unite-to-set-covid-19-recovery-pathway?utm_source=e-shot&utm_medium=email&utm_campaign= JointstatementonCovid-19recoveryFinal

IQAir (2019). 2019 World Air Quality Report, IQAir, https://www.iqair.com/world-most-polluted-cities

IDCOL (2020), Infrastructure Development Company Limited, Dhaka, http://idcol.org/ home/solar

IRENA (2020). COVID-19 and renewables - impact on the energy system, Webinar, International Renewable Energy Agency, 4 June 2020, https://www.irena.org/ events/2020/Jun/COVID-19-and-renewables—impact-on-the-energy-system

Khadka S N (2012). Asia's energy crisis demands collective action, BBC, 12 August 2012, https://www.bbc.com/news/business-19107372

MNRE (2020). Physical Progress, Ministry of New and Renewable Energy, Government of India, https://mnre.gov.in/physical-progress-achievements

PWC (2020). COVID-19: What it means for the energy industry, PricewaterhouseCoopers, https://www.pwc.com/us/en/library/covid-19/coronavirus-energy-industry-impact.html

Schwab K (2019). The Global Competitiveness Report 2019, World Economic Forum, Geneva.

UN (2020a), Nearly half of global workforce at risk as job losses increase due to COVID-19: UN Labour Agency, United Nations News, 28 April 2020, https://news.un.org/en/ story/2020/04/1062792

UN (2020b). COVID-19 Response, The Department of Global Communications, United Nations, https://www.un.org/en/un-coronavirus-communications-team/un-urges-countries-%E2%80%98build-back-better%E2%80%99

Panda (2016). South Asian Power Sector & Cross Border Electricity Trade, Power-Gen India & Central Asia Conference, 18-20 May 2016, New Delhi

Wijayatunga P and Fernando P (2013). An Overview of Energy Cooperation in South Asia, South Asia Working Paper Series, No. 19 Asian Development Bank.

WB (2020). In cooperation with IEA, IRENA, UNSD, World Bank, WHO. Tracking SDG 7: The Energy Progress Report 2020. World Bank Group, Washington DC. © World Bank. License: Creative Commons Attribution—NonCommercial 3.0 IGO, CC BY-NC 3.0 IGO.

Zhang F (2019). In the Dark: How Much Do Power Sector Distortions Cost South Asia? South Asia Development Forum, World Bank Group, Washington DC: © World Bank. License: Creative Commons Attribution CC BY 3.0 IGO; doi:10.1596/978-1-4648-1154-8

Index

A

Agriculture residue, 80, 83, 112
Air pollution, 8, 32, 105, 187
Asian Development Bank (ADB), 29, 47, 135, 234, 261
Association of South Asian Nations (ASEAN), 143, 218

B

Bagasse, 107, 112, 117, 152, 160, 176, 255
Biodiversity, 5, 10, 87, 207
Bioenergy, 114
Biofuel, 113, 128, 213, 250
Biogas, 25, 84, 112, 240, 255
Biomass
    contribution, 7, 9, 40
    emissions, 35
    energy, 112
    potential, 83, 84, 107
    resource, 3, 7, 43
Buildings, 23, 203, 211, 247
Built environment, 136

C

Carbon dioxide, 33, 87, 89, 158, 229, 232, 253
Central Asia-South Asia (CASA), 261
China Pakistan Economic Corridor (CPEC), 11, 184, 190, 261
Clean Development Mechanism, 31, 210
Climate change
    adaptation, 83, 231-238, 241, 243
    broader impacts, 2, 3, 7, 8, 12, 103, 130, 176, 207, 261
    climate action, 4
    definition, 31
    economic impacts, 10
    energy sources, 20, 64, 175
    environmental policy, 116, 127, 210
    Global Climate Risk Index, 102, 210, 223, 227, 228
    health issues, 29
    mitigation, 30, 116, 232-239
    resilience, 231
    sea level rise, 137
Coal, 44, 86, 157
Cooking stoves, 151, 248, 254
COVID-19, 6, 14, 19, 218, 259, 260, 263

D

Decarbonization, 177, 253
Decentralized energy, 113, 118
Digitalization, 254
Distributed generation, 247, 253
Drought, 2, 8, 203, 210, 243, 262

E

Ecology, 54, 112, 238
Ecosystem, 4, 31, 58, 116, 207, 235
Electricity
    access, 3, 7, 51
    consumption, 1, 7, 22
    generation, 19, 67, 199, 222
    grid, 41, 199, 213
    mix, 116, 224
    sector, 40, 54, 169, 184
    tariff, 41, 86
Electric vehicles, 162, 199
Energy
    challenges, 3, 91, 246, 250
    crisis, 65, 147, 203
    efficiency, 23, 41, 56, 203
    mix, 64, 90, 184, 224
    policy, 22, 71, 201
    price, 41, 129, 186, 253, 262
    resources, 7, 105, 253, 260
    security, 13, 56, 57, 65
    trading, 70, 72, 86
    transition, 136, 189, 253, 260
Environment
    challenges, 3, 6, 186, 250, 263, 265
    emissions, 262
    security, 13, 14, 165, 254, 262
European Union, 218, 253, 259

F

Feed-in-tariff, 86, 248
Firewood, 166, 204
Flood, 2, 103, 210, 258, 262
Food security, 114, 133, 231, 243
Fossil fuels
    consumption, 22, 31, 77, 90, 124
    depletion, 2, 52, 58, 253
    electricity generation, 58
    emissions, 30, 58, 258
    energy content, 69
    energy generation, 56, 222

energy systems, 136
imports, 74, 78, 124, 224
potential, 70
reserves, 221
supply mix, 225
types, 66, 169
Forest, 4, 23, 70, 207, 240, 261
  forestry, 10, 31-33, 35, 36
  deforestation, 31, 145, 207, 247, 258,
Freshwater, 7, 209

G

Gasification, 112
Geopolitical issues, 11, 12, 142
Geopolitics, 10, 56
Geothermal energy, 23, 102, 206
Global warming
  contributors, 33, 98, 114
  definition, 158
  global warming potential, 114, 158
  impacts, 2, 7, 98, 231, 202
  mitigation, 71, 105
Grameen Shakti, 254, 255
Greenhouse gas emissions (GHG)
  emission sources, 3, 32, 33, 53, 102, 127, 176,
    186, 223
  impacts, 3
  mitigation, 29, 133, 147
  policies, 31,
Grid-connected, 107, 109, 242
Gross Domestic Product, 5, 18, 116, 123, 149,
    183, 257

H

Heat wave, 2, 8, 176, 186, 240
Himalayan glaciers, 7, 235
Human Development Index, 2, 19, 226, 240
Hybrid systems, 23, 51, 135
Hydropower
  generation, 19, 81, 199, 221, 233
  hydroelectricity, 70, 90
  potential, 22, 80, 117, 260
  projects, 200, 240

I

Improved cooking stoves (ICS), 240, 255
Indus
  River, 11
  Valley, 5
  Water Treaty, 12, 13
International Energy Agency (IEA), 1, 98, 100,
    263
International Hydropower Association (IHA),
    260

International Renewable Energy Agency
    (IRENA), 260

L

Levelized cost of electricity, 85, 86
Low carbon
  development, 239
  energy transition, 248
  strategy, 133

M

Macroeconomic, 142, 180
Micro-credit, 254
Micro-grid, 51,
Micro hydropower, 200, 240
Millennium Development Goals (MDGs), 3
Mini-grid, 23, 135

N

Net-metering, 23, 25
Nuclear flashpoint, 11
Nuclear power
  fusion, 108
  generation, 40, 56, 91
  share, 51, 176, 179, 222, 225
Nuclear weapons, 10, 11

O

Off-grid, 19, 41, 110, 205, 254
Organization for Economic Cooperation and
    Development (OECD), 1, 142
Ozone, 26, 187

P

Paris Agreement, 104, 117, 142, 190
Particulate matter (PM), 8, 58, 187
Petroleum, 45, 66, 88, 145, 159
Photovoltaic, 23, 86, 107, 118, 182, 206
PricewaterhouseCoopers (PWC), 260
PVC, 114

R

Rainwater, 114, 232
Regional cooperation, 13, 41, 218, 260, 262
Renewable energy
  cost, 85
  policies, 30, 40, 116, 201
  potential, 19, 78, 113
  projects, 106, 240, 255
  resources, 42, 105, 188
  strategy, 26, 46, 105, 236

targets, 46, 116
technologies, 23, 25

S

SAARC
   countries, 221, 222, 226, 233, 246,
      249, 250
   region, 63, 88, 218, 224
   regional cooperation, 5, 13, 220
Sea-level rise, 130, 137
Seawater, 7
Smart meters, 76
Socio-economics, 10, 12
Solar energy
   case study, 118, 255
   installed capacity, 109
   panels, 114
   policies, 116
   potential, 83, 107, 221
   radiation, 64, 78, 108, 205
   solar home systems (SHS), 40, 255
   targets, 47, 109, 157, 232, 240
   technologies, 23
South Asia, 3, 217, 254
Stand-alone, 23, 114
Sun, 46, 60, 105, 108
Sunshine, 85, 134, 138, 188
Sustainability, 3, 20, 113, 146, 248
Sustainable Development Goals (SDGs), 3, 4, 6,
      105, 130, 148, 177, 262
Sustainable energy transition, 253, 260

T

Territorial disputes, 11, 12, 263
Transmission and distribution, 56, 150,
      180, 199, 256

Transportation, 31, 45, 113
Transport emissions, 8, 33
Turkmenistan-Afghanistan-Pakistan-India
      (TAPI), 261

U

United Nations (UN), 22, 51, 204, 231
United Nations Development Program (UNDP),
      40, 254
United Nations Environment Programme
      (UNEP), 234
United Nations Framework Convention of
      Climate Change (UNFCCC), 31, 187,
      224, 226, 231
Urbanization, 2, 47, 98, 197, 259

W

Waste to energy, 23, 46, 107, 113
Water
   wastewater, 204, 209
   water disputes, 11, 12, 263
   waterfall, 233
   water pollution, 29, 206, 207
   water scarcity, 187, 190, 258
   water security, 10, 234, 237, 261
   wave power, 102, 105, 206
Wind power
   generation, 204
   offshore wind, 86, 111
   potential, 19, 83, 107, 199
   power density, 81, 110
   projects, 206, 244, 255
   resource, 78, 110
   targets, 23
World Bank, 12, 73, 184, 254, 262
World Health Organization (WHO), 8

Printed in the United States
By Bookmasters